East-West
Agricultural Trade

About the Book and Editor

The first study to focus specifically on the economics of agricultural trade issues in centrally planned economies, this volume contains recent findings of economists who have examined the decisionmaking processes and the trends that relate to agricultural trade with the West by Eastern Europe, the Soviet Union, and China. Future prospects for agricultural trade with these countries are considered, as are the bilateral trading relations between these countries and the United States.

Specific attention is given to the import planning process as it operates within the context of the macro domestic and international economic trade-offs confronted by planners and officials. The contributors also demonstrate that concepts such as state monopoly, socialist planning, and administered state prices are important in analyzing trade with a centrally planned economy. At the same time, however, these concepts can easily become cliches unless addressed together with the many economic variables that affect or constrain actual trade behavior. Providing a multifaceted perspective, the book explores the considerations necessary to understand global agricultural trade issues.

Dr. James R. Jones is on leave from the Department of Agricultural Economics at the University of Idaho and presently is in the International Economics Division, Economic Research Service, U.S. Department of Agriculture.

East-West Agricultural Trade

edited by James R. Jones

Routledge
Taylor & Francis Group

LONDON AND NEW YORK

First published 1986 by Westview Press

Published 2018 by Routledge
52 Vanderbilt Avenue, New York, NY 10017
2 Park Square, Milton Park, Abingdon, Oxon OX14 4RN

Routledge is an imprint of the Taylor & Francis Group, an informa business

Library of Congress Cataloging-in-Publication Data
Main entry under title:
East-West agricultural trade.
      (Westview special studies in international economics
and business)
      Includes bibliographies.
      1. Produce trade.   2. East-West trade (1945–   )
I. Jones, James R.   II. Series.
HD9000.5.EI7   1986   382'.41'091713   85-17903
ISBN 0-8133-7135-X

ISBN 13: 978-0-367-00858-1 (hbk)
ISBN 13: 978-0-367-15845-3 (pbk)

Contents

CHAPTER 1
INTRODUCTION: AGRICULTURAL IMPORT DECISION PROCESSES
IN THE CENTRALLY PLANNED ECONOMIES

James R. Jones

CHAPTER 2
THE EFFECT OF THE U.S.-SOVIET
BILATERAL TRADE AGREEMENT ON THE WORLD MARKET:
IMPLICATIONS FOR UNITED STATES POLICY

Alan J. Webb and Bob F. Jones

CHAPTER 3
HARVEST FAILURES IN EASTERN EUROPE:
PLANNERS' RESPONSES AND THEIR IMPLICATIONS
FOR WORLD GRAIN MARKETS

Josef C. Brada

CHAPTER 4
IMPORT RESPONSE, FOREIGN EXCHANGE ALLOCATION
AND INCONVERTIBILITY
IN THE CENTRALLY PLANNED ECONOMIES

James R. Jones, Hassan Mohammadi,
C.S. Kim, and Joel R. Hamilton

CHAPTER 5
THE CENTRALLY PLANNED COUNTRIES' LIVESTOCK
PRODUCT AND FEED GRAIN SYSTEMS

Kenneth B. Young and Gail L. Cramer

CHAPTER 6
CHINA: AN ENIGMA IN THE WORLD GRAIN TRADE

C. Peter Timmer and James R. Jones

CHAPTER 7
AGRICULTURAL TRADE IMPLICATIONS OF COMECON
1981-1985 PLANS

Stephen C. Schmidt

CHAPTER 8
DOING BUSINESS WITH CPEs AND
MARKETING STRATEGIES FOR WESTERN EXPORTERS

Arvin R. Bunker, James R. Jones, Dennis M. Conley

Tables

Figures

Preface

In spite of the importance attached to agricultural trade relations with the centrally planned economies, empirical knowledge and understanding of CPE agricultural trade behavior is still sketchy. This book's ultimate focus is (1) to improve understanding of how import decision processes in the centrally planned economies operate; (2) to improve awareness of what forces influence their agricultural product purchases from the United States and other competing suppliers; and (3) to improve the theoretical and empirical base for agricultural policy analysis and market strategy formulation. The book is written for specialists in agricultural trade and marketing and/or centrally planned economy research, but other students of East-West relations -- be they scholars of other persuasion, policy makers, or practitioners concerned with the actual affairs of the world agricultural trade-- will hopefully find the book informative as well. To our knowledge this is the first book to focus exclusively on the economics and commercial aspects of agricultural trade issues with centrally planned economies.

James R. Jones

Acknowledgments

The editor thanks the other contributors to this book for their conscientious efforts. The book is a group effort, but we bear responsibility for errors and shortcomings individually on our independent chapters. Many other people besides the editor and other contributors had a role in this effort. Mrs. Barbara Schnabel is given the editor's special thanks for her patience and technical competence in preparing the manuscript and bringing the book to press. The editor owes his gratitude also for many helpful remarks on various parts of the book and for other assistance received from the East Europe/USSR Branch and other International Economics Division staff in the U. S. Department of Agriculture Economic Research Service.

Resources and assistance were provided to the editor by the North Central Regional Grain Marketing Research Committee (NC-160), the University of Idaho, and the International Economics Division of the U.S. Department of Agriculture.

J.R.J

1

Introduction: Agricultural Import Decision Processes in the Centrally Planned Economies

James R. Jones

In much of the post World War II era competition for minds in conjunction with military and diplomatic posturing has dominated relations between the Western market economies and the nations whose social and economic systems are organized along lines associated with the precepts of Marxism and central planning. Fortunately trade has proven to be an area where the two camps of nations have found they have common interests in conducting relations with each other. Agricultural trade has become one of the focal points of economic and political relations between the United States and the centrally planned economies in Eastern Europe, China, and the Soviet Union. Farmer and other agribusiness groups in the United States see their economic fortunes depending heavily on agricultural import decisions in these countries.

The term "East-West" as it is used in the title of this book is somewhat dated since it was originally used to describe the regional composition of nations who were diplomatic and ideological rivals. In fact, Yugoslavia broke out of the so-called Eastern bloc early. China's political alliance with the Soviets soured in the 1960s and diplomatic relations developed between the United States and China in the 1970s. Therefore, the usage of the term East-West trade takes a different context. Usage of this term to denote relations between centrally planned economies and the market economies holds over as a convenient short hand way of highlighting certain unique features of bilateral dealings between the two groups. East-West trade relations, and agricultural trade relations in particular, focus upon the unique impact of the centrally planned economies on the economic environment faced by the United States and other economies outside this group.

The centrally planned economies have assumed a major role in the international grain market. In 1960 grain supply and demand for this group was roughly in balance, but by 1980 these countries accounted for no less than a third of world imports. Typically, two-thirds or more of the total value of U.S. exports to these countries has consisted of agricultural commodities. In the 1970s they provided one of the most rapidly growing market outlets for U.S. produced farm products. The size and variability of agricultural imports in the centrally planned economies have had a notable impact on the volume of world agricultural trade and the level of agricultural prices.

The mechanisms, institutions, and policies employed by the centrally planned economies in foreign trade matters were initially devised in the Soviet Union and later adopted by the other communist bloc countries. Given the variations in cultural and historical background and economic circumstances among these countries, this common origin of the centrally planned economies' models of international trade by no means provides a stereotyped characterization of trade behavior. Indeed, most of the centrally planned economies that initially adopted the "Soviet Model" have embarked on their own unique paths of economic and institutional evolution. Nevertheless, there is still the shared denominator of all these economies subordinating import decisions to the needs of central planning and the tenets of state socialist dogma to some degree. Trade has historically been transacted through state controlled enterprises and trading organizations and this is still true. Other shared attributes also continue.

Thus, when Western firms plan strategies to transact business with these state agencies, or when Western governments consider trade policy matters with nonmarket economies, certain anomalies need to be recognized beforehand. The socialist principle of economic organization may create unique circumstances because of the existence of a state trading monopoly. Planners direct the economy. By and large prices either play a passive role or, at the other extreme, are used as policy instruments. Instances of prices being automatically determined by market forces are rare. Planners seem to prefer bilateral trading arrangements. While the power exerted by planners is considerable, this control is subject to limitations because of harvest uncertainties and hard currency restrictions. These and other considerations are examined in this book from the perspective of their implications on decisions to import from the West, and the United States in particular.

TRADE AGENCIES IN THE CENTRALLY PLANNED ECONOMIES

In this introductory chapter we will start by reviewing some of the institutional and behavioral

pecularities of the subject countries that are salient
to discussing their involvement in international agricultural markets. Economists in market type economic
systems are accustomed to simplifying their analysis by
concentrating on three sets of actors: consumers, producers, and policy makers. In so-called nonmarket or
socialist economies it is conventional to speak of one
group of actors, "planners". This would seem to sim-
plify understanding trade decision processes, but in
fact there are several echelons of decision makers who
comprise this group.

The agencies in the nonmarket economies that foreign
exporters usually deal directly with are called Foreign
Trade Organizations (FTOs) or, alternatively, foreign
trade enterprises or foreign trade corporations. For all
intents and purposes, these are the operational entities
in the State trading monopoly. However, it is inadequate to focus on the operations of these agencies in
isolation when trying to delineate how decisions are
arrived at regarding the composition, quantities, and
sources of imports. A number of other agencies or
governmental bodies are also involved. Figure 1.1 presents a typical organizational chart of bodies involved
in the import decision process. The decisions of FTOs
are influenced in varying degrees of detail by higher
supervisory and planning organs in the planning apparatus and/or by ancillary agencies charged with related
functions. Other important foreign trade institutions
to be considered include the Ministry of Foreign Trade,
the Foreign Trade Bank under the supervision of the
Ministry of Finance, and the appropriate industrial
ministry (in this study this would usually be the Ministry of Agriculture). Indirectly (and sometimes directly)
critical decisions on import questions are made in the
Central Planning organ (Gosplan in the Soviet Union, the
State Planning Commission in Poland, etc.) and/or the
Council of Ministers, or even in the central committee
or party congress where very broad priorities are often
established (for example, to stress improving the diets
of the populace by increasing livestock production over
the course of a five-year plan).

The role of enterprise end-user managers in import
decisions varies from country to country and by sectors
or commodities. Evidence suggests that the trend in
varying degrees is towards granting more authority to
end-using enterprises in import decisions. Recognition
of the importance of giving end-user enterprises more
direct access to the source of their imports has led
Hungary, the German Democratic Republic, Bulgaria, and
particularly Yugoslavia to permit direct negotiations by
certain of their domesic end-user enterprises. Still
direct end-user imports, without the mediation of the
FTO, are rare. In certain cases the end using enterprises are allowed to retain part of the hard currency

4

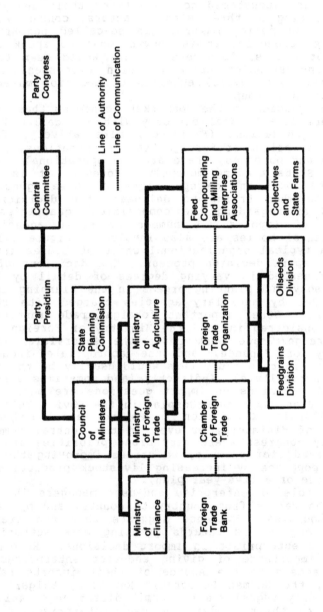

Figure 1.1 Foreign Trade Structure of a Centrally Planned Economy

Party
Congress

Central
Committee

Party
Presidium

——— Line of Authority

········· Line of Communication

State
Planning
Commission

Ministry
of
Agriculture

Feed
Compounding
and Milling
Enterprise
Associations

Collectives
and
State Farms

Council
of
Ministers

Ministry
of Foreign
Trade

Foreign
Trade
Organization

Feedgrains
Division

Oilseeds
Division

Chamber
of Foreign
Trade

Ministry
of
Finance

Foreign
Trade
Bank

earned from their exports to meet their own needs for imported items. However, to date this practice has had little impact on large-scale imports of grains and oil-seeds. In the more usual case where the FTO acts as the agent in contact with the foreign exporter, end-using enterprise managers ultimately can have significant indirect if not direct influence in the decisions of what to purchase. Even if direct decisions are not made by enterprise managers on imports, decision makers at higher levels depend on them for information regarding import needs.

Certain other organizations possess trading rights in addition to state foreign trade organizations or enterprises. Considerable reorganization of foreign trade activities occurred in selected Eastern European countries in the late 1960s and early 1970s. These reforms in certain instances resulted in a transfer of trading rights and activities to industrial ministries and/or certain large end-using production enterprises. More recently, in China, certain trading rights were transferred to provincial level trading organizations. Since this reorganization has only occurred fairly recently, or is still in process, it is impossible to generalize the nature and extent of responsibilities and authority that has been shifted. Indeed concerns about controlling use of scarce foreign exchange has led to retraction of some of the moves to decentralize import decisions in China.

Financing of international transactions is handled by the foreign trade bank. Responsibilities of these banks are essentially equivalent to those of merchant or commercial banks in market type economies. In addition they can be useful in helping a seller identify end-using enterprises and/or FTOs possessing the necessary currency and responsibility to purchase specific items since the foreign trade bank is responsible for most foreign currency allocations. The importance of foreign currency allocations will be discussed in greater detail later in this book.

A full treatment of how foreign trade planning and operation is implemented would require a detailed trea-tise exceeding the scope of this exposition. Since sub-stantial variation exists in how planning is implemented among countries and commodity groups, it is important to emphasize two points. The import process is not simply a one-way vertical procedure with higher planning organs handing directives down the chain of command from the State Planning Commission to the Ministry of Foreign Trade, Foreign Trade Organization and production agency, regarding how much or what to import from which source. Rather, several rounds are made in formulating and implementing the foreign trade plan with lower organs having input in terms of registering needs and making checks of the consistency of foreign trade plans.

Viewed from the top echelon of the planning apparatus the process may be one of responding to pressures from lower levels as much as exerting pressures on lower levels of the economy. Secondly, many decisions regarding imports are made that deviate from the guidelines of the plans. The plan is of necessity flexible enough to allow for adjustments of imports during the planning year or five-year process when unforeseen circumstances such as crop shortfalls necessitate imports to meet certain priority needs or to prevent serious production bottlenecks or abnormal consumption shortages. Practically speaking, import allocation may be determined in an ad hoc manner through a process of negotiation between end-users, ministry officials, and higher level planning officials in reaction to or anticipation of unexpected events, as much as by predetermined planning.

INCONVERTIBILITY, BILATERALISM AND
FOREIGN EXCHANGE SHORTAGES

Balance-of-payments deficits have always plagued the nonmarket economies,[1] but in recent years shortages of hard currencies have assumed an importance in the socialist economies trade relations with the West second only to their political relations. Over full-employment growth policies plus an inability to produce goods of a quality that attract foreign buyers have historically created balance-of-payment pressures in the communist countries. Increasing demand by the populace for improved diets and more and better consumer goods produced in the West, plus a need to rely upon imports of Western equipment and technology to bolster economic development objectives, have resulted in rising import requirements. In the case of foodstuffs, and meats in particular, demand pressures due to rising incomes have been reinforced by prices maintained at artificially low levels through the mechanism of state subsidies. Shortfalls in agricultural production due to adverse natural causes and development programs that have failed to yield expected results have compounded the pressures for imports.

Analyzing the true payments and financial position of the communist economies is difficult as there are no clear cut criteria that can be used to explicitly define "how much" debt a country can handle in external trade dealings. Fundamentally, this would bear on the real resource endowments and productivity of the country in question, but this is not easily expressed in a way allowing a quantitative indication of where a nation stands in relation to its debt position. Complicating the task even further is the fact that most of the communist countries refuse to reveal the extent of their

holding of foreign hard currency reserves. Thus, no
single fully satisfactory index is available to directly
measure external debt relative to a country's financial
situation. Notwithstanding this problem, some indication of the ability of the nonmarket block to continue
to finance increases in imports of agricultural and
other goods can be obtained from looking at the net
trade position and level of indebtedness incurred
through trade credits and other forms of credit.

The Soviet bloc countries incurred balance of trade
deficits continually through the 1960s and 1970s (Table
1.1). Bulgaria, Czechoslovakia, German Democratic
Republic, Hungary and Romania failed to record a surplus
for a single year over the period of the sixties and
seventies. Poland actually achieved a surplus in three
years during this period, but the country's trade position literally hemorrhaged during the years 1973-1981.
The region as a whole witnessed an accumulated trade
deficit approaching 64 U.S. billion dollars for the
years 1960-1981.

Even by the end of the 1960s, Eastern Europe's hard
currency debt to the West was viewed to be increasing at
an unsustainable pace (Snell, 1974). Nevertheless
Soviet and Eastern European debt continued to increase
from 5.5 U.S. billion dollars in 1971 to 71.9 U.S.
billion dollars in 1981 (Table 1.2). Debt service
ratios that express interest and principal repayments on
foreign debt as a ratio of exports illustrate how the
hard debt carrying capacity of the Soviet bloc deterio-
rated in the 1970s.[2] Debt service ratios for Eastern
Europe and the Soviet Union in this period are shown in
Table 1.3. International bankers often use 25 percent
as a rule of thumb of an upper safe limit for a country's debt service ratio. Nevertheless, rapidly expanding credit by Western traders permitted continuation of
the increase in imports in the 1970s. The debt-service
ratio for the grouping as a whole climbed to 41 percent
in 1979, as compared to slightly less than 20 percent in
1972. Poland's position in particular drew alarm as its
principal due and interest payments constituted an estimated 92 percent of its exports in 1981. When in 1981
Poland had to reschedule its interest and debt payments
to its creditors, and Romania also encountered a credit
crisis, concern mounted that many of the other centrally
planned economies harbored similar problems (Journal of
Commerce, December 28, 1981). Net debt for the Soviet
bloc economies peaked in 1981 and has shrunk slightly
primarily due to a reduction in imports. The point is
that hardcurrency shortages have plagued these economies
continually. Even in the case of China, which has
eschewed accumulating large debts, it is generally
acknowledged by Chinese and foreign analysts that hard
currency availability has greatly constrained import
behavior.

TABLE 1.1
Soviet Bloc Net Trade Deficit with the Industrial West.*

Year	Bulgaria	Czecho-slovakia	GDR	Hungary	Poland	Romania	Soviet Union	Total
				--millions	of U.S.	dollars--		
1960	23.7	34.1	55.5	49.3	105.2	3.5	140.3	411.6
1961	16.9	50.7	59.4	62.9	141.5	39.2	77.4	448.0
1962	13.1	58.4	37.0	40.8	79.6	72.0	193.4	494.3
1963	17.7	23.7	3.9	44.2	50.7	57.2	245.3	441.9
1964	89.8	78.1	31.8	92.0	65.4	88.6	282.9	728.6
1965	109.5	109.4	88.9	66.0	-37.4	103.0	-207.0	232.4
1966	208.6	152.6	117.8	84.8	24.9	141.2	-209.3	520.6
1967	176.1	57.5	94.6	83.6	78.9	309.6	-320.5	479.8
1968	138.8	100.1	10.7	96.8	110.9	317.2	-100.2	674.3
1969	98.0	85.0	11.0	55.0	81.0	293.0	131.0	754.0
1970	108.2	120.0	89.6	179.7	-60.4	206.2	47.0	690.3
1971	98.0	159.9	144.8	277.5	-15.9	144.2	-77.0	731.5
1972	101.0	138.2	187.9	194.5	308.0	215.4	702.0	1,847.0
1973	154.7	185.2	230.4	200.1	1,285.5	217.0	960.0	3,232.9
1974	472.1	329.6	227.9	811.6	1,966.1	512.0	-122.0	4,197.3
1975	742.6	399.6	313.5	596.6	2,622.2	476.0	4,508.0	9,658.5
1976	538.2	537.1	466.8	597.6	2,257.8	215.0	4,170.0	8,782.5
1977	463.4	395.2	286.4	851.7	1,515.6	533.0	2,423.0	6,468.3
1978	582.8	357.1	403.7	1,334.1	1,578.7	788.0	2,935.0	7,979.4
1979	479.4	286.3	1,142.9	857.5	1,401.1	502.0	1,724.0	6,393.2
1980	838.1	31.8	849.8	977.9	1,299.6	127.0	53.0	4,177.2
1981	1,092.7	-103.4	766.6	1,198.5	1,042.5	0.0	346.0	4,342.9
1982	887.2	-278.9	-265.7	254.8	208.2	-1,017.0	-276.0	-487.4

*Negative signs denote net trade surplus.
Source: International Monetary Fund. Direction of Trade, various issues.

TABLE 1.2
Eastern Europe and Soviet Union Net Hard Currency Debt to the West (U.S. Billion $).

Year	Bulgaria	Czecho-slovakia	GDR	Hungary	Poland	Romania	Soviet Union	Total
1971	.7	.2	1.2	.8	.8	1.2	.6	5.5
1972	.9	.2	1.2	1.0	1.2	1.2	.6	6.3
1973	1.0	.3	1.9	1.1	2.2	1.5	1.2	9.2
1974	1.4	.6	2.6	1.5	4.1	2.5	1.7	14.4
1975	2.3	.8	3.5	1.2	7.4	2.4	7.5	25.1
1976	2.8	1.4	5.0	2.9	10.7	2.5	10.1	35.4
1977	3.2	2.1	6.2	4.5	13.5	3.3	11.2	44.0
1978	3.7	2.5	7.5	6.5	17.0	4.5	11.2	52.9
1979	3.7	3.0	8.4	7.3	20.0	6.7	10.2	59.3
1980	2.7	3.6	11.8	7.5	24.5	9.2	9.3	68.6
1981	2.2	3.4	12.3	7.0	24.7	9.8	12.5	71.9
1982	1.7	3.3	10.7	6.6	23.8	9.4	N.A.	--
1983	1.4	2.7	9.1	6.8	26.4	8.8	9.6	64.8
1984	.9	2.5	7.7	6.1	26.8	6.8	4.1	54.9

Sources: (1971-1979) C.I.A. National Foreign Assessment Center. Estimating Soviet and East European Hard Currency Debt, ER80-10327 (June, 1980), p. 7; (1980 Eastern Europe) C.I.A. National Foreign Assessment Center. Handbook of Economic Statistics 1982 A Reference Aid, CPAS82-10006 (September, 1982), p. 54 and 77; (1980 Soviet Union) Joint Economic Committee, Congress of the United States. Soviet Economy in the 1980s: Problems and Prospects, Part 2, Selected Papers (December 31, 1982), p. 490; (1981-1984) U.S.D.A. Eastern Europe Outlook and Situation Report, RS-85-7 (June, 1985), p. 18. Total debt for 1983 and 1984 was published in The Journal of Commerce, June 18, 1985, p. 4a from data released by the Vienna Institute for Comparative Economic Studies.

TABLE 1.3
Eastern Europe and Soviet Union Estimated Debt Service Ratios.[a]

	1972-1975 (average)	1976	1977	1978	1979	1980	1981	1982	1983
Bulgaria	34	39	45	46	38	33	34	27	22
Czechoslovakia	12	15	17	20	22	22	20	19	18
GDR	21	29	38	49	54	44	52	52	46
Hungary	17	21	25	36	37	31	33	33	31
Poland	22	42	59	79	92	97	175	215	245
Romania	24	18	19	20	22	26	27	45	32
Soviet Union[b]	18	25	27	31	24	15	17	16	17

[a](1972-1979) Repayments of principal on medium- and long-term debt and of interest on all debt as a percentage of merchandise exports to the developed West; (1980-83) same as 1972-79 for Eastern Europe but for Soviet Union includes sale of arms and gold.
Source: (1972-1979) C.I.A. National Foreign Assessment Center, Estimating Soviet and East European Hard Currency Debt, ER80-10327 (June, 1980), pp. 16-29; (1980-1983) Ronald Miller and Dennis Barclay. "Eastern Europe: Facing Up to the Debt Crisis," March, 1984, Table 13, Eastern Europe; C.I.A. Handbook of Economic Statistics, CPAS 85-10001, September, 1985, Table 48, p. 73.

A characteristic of the communist countries' foreign trade is that, without exception, their currencies are inconvertible in international transactions. In the environment of world markets, this means that the non-market economies' currencies are not accepted in international trade transactions. Also importers cannot freely convert domestic currencies into foreign exchange. Imports require an explicit appropriation of hard currency from government authorities. Furthermore, official exchange rates expressing their currencies in terms of hard currencies cannot be construed as meaningful price relationships.

Inconvertibility is both a product and a cause of extreme balance-of-payments pressures. The inconvertible status of the nonmarket countries' international financial position has important implications in that it hinders trade with foreign partners, creates pressures for bilateral as opposed to multilateral trade and payments arrangements, and generally creates a nontariff instrument to regulate imports via hard currency appropriations.

Bilateral trade and payments arrangements play an important role in nonmarket economy trade activities. Faced with inconvertibility and payments pressures, the centrally planned economies have tended to seek bilateral trade and payments arrangements. [3] Consequently, intergovernmental trade agreements, sometimes based on barter type arrangements, are common between the nonmarket economies and other trading countries. One such agreement between the Soviet Union and Argentina involved at least one million tons of Soviet cereal grain purchases; fourteen thousand tons of beef, and three million gallons of wine being exchanged annually for $300 million worth of power plant generators and oil industry equipment. Larger volume grain agreements have been signed since the U.S. embargo on the Soviets in 1980 but the extent of the reciprocal payments nature of these is not known. Hungary has had bilateral accords with Brazil and Peru providing for an exchange of engineering industry equipment, vehicles, machinery, and other items for fishmeal, coffee, and other goods. Poland and Brazil have a general agreement valued at $2.5 billion between 1978 and 1990 with Brazil shipping ore in exchange for Polish metalurgical coal (Sea Trade). In 1985 Poland arranged for lamb imports of 18,000 MT from New Zealand on a barter basis and rye was bartered against pork from the Federal Republic of Germany. Agreements with India in the past have involved India exchanging peanut meal for fertilizer. An agreement with the Peoples' Republic of China (PRC) and Peru involved a commitment by the PRC to furnish Peru 75,000 metric tons of rice and purchase 40,000 metric tons of fishmeal. The Chinese also have had an arrangement to exchange 662,000 barrels of light low sulphur oil worth

ten million dollars for 275,000 tons of Brazilian ore.
Bulgaria entered into an arrangement with Brazil to
acquire products including soybean meal in exchange for
soda ash, steel products, and pharmaceutical raw materials. Bilateral government arrangements are also important in trade dealings with developed countries, as witnessed by the grain agreements between the United States
and the Soviet Union and the Peoples' Republic of China,
along with similar Canadian Argentine, and Australian
pacts. However, as a rule, reciprocal purchases have
not been built into these agreements.

Bilateral arrangements involving individual export
firms in the West agreeing to take payment in reciprocal
purchases rather than receiving hard currency payment
are important also. Barter transactions at this level
are not extremely common in the literal sense of only
two trading parties being involved in an exchange of
merchandise for merchandise with no money being
involved. If more than two parties are involved,
"switch transactions" are on occasion arranged with the
seller taking payment in the form of merchandise that is
disposed of through a third party frequently referred to
as a specialist switch dealer. Both of these forms of
trade arrangements are relatively insignificant, but a
variety of arrangements referred to as countertrade
deals have become especially prevalent in transactions
with the Eastern European countries. Countertrade transactions occur in several forms including counterpurchase deals and product buyback or compensation deals.
Counter-purchase is the most prominent form of countertrade arrangement. In this case the Western exporter
agrees to buy or import products equivalent to some percentage of his sales. The Western supplier usually
receives full cash payment but he agrees to buy products
from the Eastern firm in a separate but linked contract.
Counterpurchase deals ordinarily allow the Western
exporter time (usually one year) to select and purchase
the specified goods in return. In a compensation or
buy-back arrangement, the Western firm provides plant
and/or equipment and technology with the agreement that
payment or partial payment will be received by the
exporter taking part of the resultant output in return.
The intracacies and implications of countertrade transactions are discussed in greater detail in Chapter 8 in
this book.

The fact that inconvertibility and a lack of hard
currency leads to bilateralism has implications for the
import capacity of the countries in question. As will
be emphasized later in the book, inconvertibility suggests that hard currency allocations may be an explicit
consideration in determining import response to world
price movements and domestic crop shortfalls. It is
also important to note that, given inconvertibility,
bilateral arrangements theoretically may allow more

trade than would otherwise occur. Still, the level of
trade can seldom reach as high a level, nor claim all
the economic rewards of international specialization
that could be reaped in an environment of convertible
exchange and multilateral transactions.[4] Several
Eastern European economies in particular recognize
this. Reforms were instituted in the 1960s and early
1970s to modify their internal pricing mechanisms so as
to reduce or eliminate the insulation of their domestic
price movements, which is a necessary precondition of
convertibility. These reforms have met with little success to date, however. In the words of Bila Csikos-
Nagy, the President of Hungary's Price and Material
Office, "Adjustment to the new relative world market
prices has proved to be a much graver problem, consuming
much more time than had been assumed," [sic] (Csikos-
Nagy 1979). The evidence is not encouraging that they
will succeed in the foreseeable future (Bornstein).
Experience has shown it to be extremely difficult to
effectively establish decentralized market planning and
national pricing schemes, which are the conditions
necessary if they are ever to rid themselves of inconvertibility.

Other reasons behind the communist nations' inability to export sufficient amounts to finance their growing import needs can be traced in some measure to their
own foreign trade institutions and behavior. As noted
by Brada and Jackson (1978), a weakness of a planned
economy is its inability to link the producer to the
world market in a meaningful way. Linking the activity
of domestic enterprises directly and automatically to
world market developments is inconsistent with detailed
central planning of domestic activity. Export producing
enterprises in the past have lacked incentive to produce
for export, or to meet the special requirements of
foreign buyers, because they responded to quantitative
targets that were usually not differentiated among
buyers (East-West Markets). Managers may rationally
understate their export capabilities in order to keep
export targets low. Furthermore, the profits of individual enterprises have not been dependent upon exports.
Given a policy of taut employment planning, where the
enterprises may be able to sell all they produce domestically, the incentive to find export market outlets is
lacking. The ingenuity of the enterprise managers has
been focused upon attaining inputs, not acquiring access
to new markets overseas. To the extent that exported
goods require special attention, they make it more difficult for the enterprises to satisfy their assigned
targets. Where this applies, there is actually a negative incentive to produce for export. Also, there has
been little direct communication between producing
enterprises and overseas buyers since trading agencies
assume this role. These problems, along with numerous

others, have undoubtedly contributed to an unfavorable reputation for the centrally planned economies' manufactured goods in Western markets.

The authorities in the centrally planned economies are not blind to the difficulties noted above. It was noted earlier that reforms in Hungary and other Eastern European governments, and more recently in China, have focused upon decentralizing decision making and instituting increased incentive schemes in individual enterprises. If reforms do not correct the problem of low-quality production of export goods, the ability of the centrally planned economies to generate hard currency through exports will continue to suffer. This in turn will hinder their ability to procure imports from the West.

The reputation for poor product quality is by no means the only deterent to penetration of hard currency markets. A legacy of "politicization" will undoubtedly continue to impair trade relations with the Soviet Bloc. Denial of most favored nation (MFN) treatment by the United States to the Soviet Union, the German Democratic Republic, Bulgaria and Czechoslovakia, and retraction of this status to Poland following the declaration of martial law, still remains as a potential obstacle, though empirical work on this subject leaves room for considerable speculation on how much impact discriminatory treatment of these countries' exports has on East-West trade, and particularly agricultural trade.

An important point to bear in mind from the above discussion is that the tendency to assume that authorities in the centrally planned economies wield unlimited power to control imports can be misleading. Hard currency shortages constrain the scope of their discretion over import decisions. There has been a tendency to assume that the planners are inflexible once they have set their targets, but the necessity to cope with hard currency shortages has restricted this power. This has been especially true since the late 1970s. We will next look at the role of prices in import decisions, and in a later chapter hard currency allocation and import response will be explored in more detail.

ROLE OF PRICE IN IMPORT DECISIONS

Empirical evidence documenting the response of import decisions in the centrally planned economies to world price movements is extremely tenuous as is noted in Chapter 4. Some analysts have assumed that world price movements have no effect in these economies. At the opposite extreme interviews with trade officials in the centrally planned economies and with Western grain traders indicate that the selection of the source of imports hinges upon which prospective seller offers the

TABLE 1.4
Retail Prices of Selected Foodstuffs in the German Democratic Republic.

	Unit	1960	1965	1970	1975	1979	1983
				Marks/Unit			
Potatoes, excluding early potatoes, packaged	5 kg.	0.60	0.85	0.85	0.85	0.85	0.85
Wheat flour, superfine, in 1,060 gram packages	kg.	1.32	1.32	1.32	1.32	1.32	1.32
Rye-mix bread	kg.	0.52	0.52	0.52	0.52	0.52	0.52
White bread	kg.	1.00	1.00	1.00	1.00	1.00	1.00
White sugar, packaged	kg.	1.50	1.50	1.55	1.55	1.55	1.55
Beef roast, without bones	kg.	9.80	9.80	9.80	9.80	9.80	9.80
Pork cutlet	kg.	8.00	8.00	8.00	8.00	8.00	8.00
Sausages	kg.	6.80	6.80	6.80	6.80	6.80	--
Whole milk, 2.5% fat content, in bottles	liter 1/2 liter	0.36	0.36	0.36	0.36	0.36	0.68
Butter, in 250 gram packages	kg.	10.00	10.00	10.00	10.00	10.00	--

Sources: Agricultural Attache Report No. GE9004, January 29, 1979; No. 5Y1002, February 19, 1981, U.S. Embassy in East Berlin; and Statistical Yearbook of the German Democratic Republic, 1984.

TABLE 1.5
Official Retail Prices of Major Products in Peking Under
Planned Distribution (Yuan/kg.)

Year	Rice Average or Standard	Rice Top Grade	Wheat Flour Ordinary or Standard	Millet	Cotton Cloth (Yuan/Shih Chih) Wu-fu or Kung-nung
1952	--	0.39	0.37	0.22	0.295
1953	0.296	0.414	0.368	0.26	0.285
1954	0.296	0.414	0.368	0.26	0.285
1955	0.296	0.414	0.368	0.26	0.285
1956	0.296	0.414	0.368	0.26	0.285
1957	0.296	0.414	0.368	0.26	---
1958	0.296	0.414	0.368	0.26	---
1959	0.296	0.414	0.368	0.26	---
1960	0.296	0.414	0.368	0.26	---
1961	0.296	0.414	0.368	0.26	---
1962	0.296	0.414	0.368	0.26	---
1963	0.296	0.414	0.368	0.26	---
1964	0.296	0.414	0.368	0.26	---
1965	0.296	0.414	0.368	0.26	---
1966	---	0.40	0.32-0.36	---	---
1967-1970	--------------------Not Available------------				
1971	---	0.40	0.36	---	---
1972	---	0.40	0.33	---	0.28
1973	---	---	---	---	0.28
1974	---	0.41	---	---	0.28
1975	---	0.42	---	---	0.28
1976	---	0.40	---	---	---
1977	---	---	0.37	---	---

Source: Tong-eng Wang. Economic Policies and Price Stability in China, Institute of East Asian Studies
University of California, Berkeley, Center for
Chinese Studies, China Research Monograph No. 16,
1980, pp. 14-15.

Table 1.6 Average Prices Received by Producers for Grains.[a]

Year	Czechoslovakia (Kcs/100 kg)	GDR (Marks/100 kg)	Hungary (Forints/100 kg)	Poland (Zlotys/100 kg)	World Index for Grain (1952-56=100)
1962	124.16	33.39	264.00	221.58	87.6
1963	124.58	32.65	247.20	233.71	87.5
1964	132.60	36.43	261.40	235.66	87.8
1965	125.90	37.68	305.00	282.03	88.8
1966	151.60	38.58	308.00	280.49	94.9
1967	156.20	38.23	331.40	281.96	94.6
1968	155.80	38.60	366.40	297.56	89.7
1969	154.40	38.30	305.00	302.12	87.6
1970	158.20	38.21	293.80	287.55	90.5
1971	160.80	39.05	317.00	312.40	84.1
1972	165.20	38.87	306.60	343.60	110.9
1973	166.00	37.51	287.00	351.00	206.5
1974	164.40	37.71	304.80	359.60	250.5
1975	165.00	40.04	296.80	364.80	180.8
1976	162.60	38.36	321.40	496.20	150.9
1977	163.00	38.85	327.60	485.00	149.6
1978	164.60	38.96	331.80	505.00	175.3
1979	164.20	37.73	341.60	477.75	216.9
1980	166.60	37.59	358.00	467.00	256.4
1981	167.60	38.64	362.80	974.25	212.5
1982	177.20	40.67	361.00	1534.00	192.0

Coefficient of Variation:

	Czechoslovakia	GDR	Hungary	Poland	World
1962-1982	9.9	4.9	10.7	64.9	42.6
1962-1972	10.9	5.9	11.3	13.1	7.9
1973-1982	2.5	2.8	8.2	62.0	18.5

Trend P=132.2+2.14 year P=35.8+.187 year P=272.07+3.83 year P=53.91+34.46 year P=54.94+7.97 year

Correlation with World Price: .637 .280 .449 .479

a Averages of monthly rye, barley, wheat, corn, and oats.
Source: U.N. (ECE). 1959-1983.

most favorable price terms, other things equal. It is
argued in Chapter 8 that, barring a long term trade
agreement or a countertrade arrangement, price competitiveness is a critical determinant of making a sale,
although service is important too. While there is
little doubt that the authorities are very responsive to
who gives the best price bid, there is still the issue
of the effect of world price movements on the overall
level of imports.

The consensus among many analysts is that imports
are not highly responsive to world price movements where
centrally planned economies are involved. It is generally argued that these economies tend to insulate domestic prices from world price movements. Before profits
and prices began to assume an active role in the socialist bloc countries, physical targets and allocations
dictated what was consumed and produced. In the case of
domestic retail prices, there was little movement. For
example, Tables 1.4 and 1.5 show that in the German
Democratic Republic retail prices of basic agricultural
products remained basically unchanged from 1960 up
through the 1980s, and in China the same was true prior
to the post-Mao economic reforms. Today, China and most
of the Eastern European economies have begun to abandon
the policy of fixed retail prices.

Historically, CPE authorities have always been somewhat more flexible in adjusting prices received by agricultural producers than prices paid by consumers. Time
series data available on Czechoslovakia, German Democratic Republic, Hungary, and Poland on prices received by
producers for grain, reveal coefficients of variation of
2.5, 2.8, 8.2, and 62.0, respectively, for the years
1973- 1982 (Table 1.6). In terms of world price movements, three of these economies show less price variation than the world index of grain prices, although
Poland surprisingly shows more. The data show a generally positive trend in these prices and world grain
prices, but their degree of correlation with world price
movements is modest.

In truth all that we can conclude is that retail
prices have been sticky in the past and producers'
prices have been less rigid. Besides the fact that the
above data do not cover all countries, there is a lack
of proper domestic deflators and comparable international deflators for a systematic evaluation of the
relative behavior of world and domestic prices. For
example, in the Soviet Union the total index of procurement prices for agricultural production increased 1.62
times between 1964 and 1977, but costs per unit of output increased 2.1 times so real prices actually declined
(Markish, p. 217). Thus, looking at nominal procurement
prices tells us little about time price behavior.
Exchange rate considerations also cloud such comparisons.

Timing complicates the question. In response to domestic and international budget pressures in the late 1970s and the early 1980s, several price moves have been introduced in the CPEs. In the case of Poland, only modest change occurred in the domestic food price index in the 1970s, but prices were increased abruptly in the 1980s (see Table 1.7). Social upheavals in Poland attest to the costs associated with such radical price movements, but the predicament that confronted the decision makers was that due to the long delay, abrupt price changes were required. Similar price announcements have been made in the other Eastern European economies, but again it is difficult to relate this action statistically to world price movements.

World price increases in the period 1972-1975 may have been indirectly a factor stimulating these price initiatives that occurred in the late 1970s and early 1980s, but this may be difficult to verify or deny given the delayed and erratic price policies initiated only after a crisis forced such action. Research on CPEs and the role of prices has only looked at these issues superficially if at all. Chapter 4 attempts to address the issue in a conceptual framework. Suffice it to say at this point that the role of world prices in import behavior is a complex issue which needs to be better addressed. Part of the answer lies in the means by which world prices are transmitted to domestic prices. It will be argued in Chapter 4 that we also have to know more about the actual process of allocating foreign exchange.

Table 1.7 Food Price Indices in Poland (1970=100)

	1970	1975	1978	1980	1981	1982	1983
Socialized Mkt.	100	104	120	143	165	431	485
Private Mkt.	100	137	204	281	435	775	777

Source: Compiled from Maty Rocznik Statystyczny 1984 Warsaw p. 252 by David B. Houston, "Poland's Economic Crisis" Eastern Economic Journal, Vol. X, No. 4 (October-November, 1984). p. 448.

EAST-WEST AGRICULTURAL TRADE ISSUES

With this brief introduction to the institutional and economic setting of the import decision process of the centrally planned economies, we now turn to the essays in this book and briefly introduce the topics they will address. The emergence of the centrally planned economies as important purchasers of agricultural products on the world market has raised questions for traders, policy makers, and economists which center on three related issues. The first issue concerns the implications of the variability and unpredictability of this group of buyer's agricultural purchases on the economic welfare of producers, exporters, and consumers in the world food system. The second issue is how to predict the volume (and timing) of grain purchases by the socialist economies. From the seller's viewpoint there is a third issue: how does one do business with these buyers?[5] While all these issues are related we will discuss them separately for expositional convenience.

LONG-TERM GRAIN AGREEMENTS

Earlier it was noted that socialist planned economies tend to conduct business on a bilateral basis in order to ease their hard currency problem. There is another motive for signing bilateral agreements. Planners use bilateral trade agreements to assure access to supplies of goods considered essential to the success of economic plans.

Economic considerations relating to long-term trade agreements are complex, as demonstrated in the two chapters (Webb and Jones, Brada) in this book treating the subject. It is important to point out that neither of these studies claims that it considers all aspects of trade relations as they are affected by long-term bilateral agreements. International diplomacy goes beyond the scope of the studies presented in this book. Historical East-West rivalry is a major factor in trade relations. The United States-Soviet long-term agreements in 1976 and 1983, may have served the purpose of "partially insulating the grain trade from the variable political relations between the two superpowers" (Hardt 1983). The January 4, 1980 embargo on grain exports to the Soviet Union in response to the Soviet invasion of Afganistan was placed only on the seventeen million tons of grain offered to the Soviets in excess of the contracted eight million tons negotiated in the 1976 grain agreement. On the other hand, long-term agreements are not inviolable as witnessed by the Chinese refusing to meet their committed purchases in the 1983 textile dispute with the United States. On a positive note however, these agreements are themselves a part of a process of diplomatic negotiation that at least signifies some level of willingness between political powers to normalize their trade relations.

In this volume long-term agreements are viewed in terms of their usefulness as a Western export instrument to impose greater stability on international grain prices and to effect more favorable terms of trade vis a vis the socialist state trading monopoly.

A common perception is that a centrally planned state trader intrinsically possesses greater potential to exercise market power than a market economy which conducts trade through its private sector (McCalla 1981). It is intuitively easy to arrive at such a conclusion. Typically these economies have operated on explicit recognition of the principle of a foreign trade monopoly. This may give the trading entity an adequate share of the market to exercise its power and its power could be further enhanced by virtue of its wielding superior information vis a vis its private trading partners.

The second portion of Brada's article goes to the heart of this question with a rigourous trade offer curve analysis. He demonstrates that the ability of the centrally planned economies to assert economic power through an optimal tariff formula depends upon the assumed elasticity of their trade offer curves. Brada points out that if trade insensitivity on the part of the centrally planned economies exists, the United States actually would hold the upper hand in its potential to assert its economic power. Brada demonstrates how long-term agreements with negotiated levels of trade can make it easier for the United States to exercise its market power. Even if the state traders don't necessarily wield an advantage in economic power per se, they can extract gains through secrecy. On this count Brada concludes that long-term trading agreements are potentially valuable tools to Western traders in that these agreements force the state trader to show their time-trade intentions if they wish to procure grain in excess of the minimum level guaranteed by the United States in the agreement.

One consequence of the variability and unpredictability of imports by the planned economies in the 1970s was that analysts, policy makers, and traders became interested in how long-term agreements affect world price stability in grains. Under what circumstances do they reduce variability? Alternatively, do they actually make the grain trade and prices more unstable?

The article (Chapter 2) by Webb and Jones stresses that whether bilateral agreements reduce trade and price variability depends upon the circumstances prevailing at the time. They argue that in the event of a major crop shortfall in the exporting country, importing countries locked into these agreements could be required to purchase unwanted quantities at the associated higher prices and that consequently world prices could be forced to higher levels than they would have been in the absence of the agreements. The agreement prevents the full transmission of world market price changes onto domestic markets. On the other hand, if the importing country has a large crop and lives by the terms of its agreement, there could be a smaller fall in world prices and quantities traded than if no agreement existed. Bilateral agreements can thus increase or decrease instability depending on the source of changes in world markets. Obviously the effect of a bilateral agreement also depends upon whether the agreement is honored or not . Transshipping of grain or reneging on the commitments, as the Chinese did, also has a bearing on the effect of these agreements. At any rate, the authors conclude that a bilateral agreement with a large centrally planned trader such as the Soviet Union could help stabilize world prices by preventing domestic fluctuations in supplies from being fully translated into transactions on world markets.

PREDICTING AGRICULTURAL PURCHASES BY THE CPE GROUP

Planners' behavior becomes the focal point of predicting agricultural imports in centrally planned economies. It was pointed out above that pinpointing the relevant group or agency that comprises this group is more complex than the term at first seems to imply. There are many echelons of decision making groups who interact in planning and executing imports of agricultural products. Moreover, the make-up of this group and the policy parameters and lines of authority that mold their behavior vary by country and over time. While each country has adapted its planning and economic system to its own unique circumstances under the influence of its peculiar historical setting, similarities prevail that contrast import behavior to that in market type economies. As Brada points out (Chapter 3), import decisions are made by many independent agents in market economies reacting to price signals. He argues that, in contrast, in the socialist economies the law of large numbers does not prevail and import decisions are executed less in response to price signals and more on the basis of policy decisions of government planners.

An economic reality that has to be confronted ultimately by centrally planned economies is how they are going to pay for their imports. When world prices go up the authorities have to make choices on whether or not to maintain their level of imports. When prices go down, other things equal, they can afford more. Jones, et.al., focuses in Chapter 4 on theoretical and empirical modeling issues as they bear on the relationship between world price movements and foreign exchange allocation decisions on the one hand and agricultural import response on the other. There is much work that needs to be done in this area to improve our ability to predict agricultural behavior in these countries. This chapter does not attempt an actual forecast of future imports, but it attempts a step in this direction by looking at the implications of hard currency shortages, inconvertibility, and pricing behavior discussed earlier in this introduction for modeling and explaining import behavior in the centrally planned economies.

Young and Cramer (Chapter 5) stress the importance of the policy decision in the Soviet bloc economies in the 1970s to upgrade livestock production and consumption and sustain the program with massive feedgrain imports. Using an input-output formulation, Young and Cramer investigate interrelationships between the centrally planned economies' feed and livestock sectors and on this basis project long-term import needs of this group of countries through the remaining part of the twentieth century. As they note, these projections hinge upon several "ifs" including whether planned production achieves certain target levels and whether the foreign exchange capabilities of these countries will sustain import needs. The projection technique is especially interesting to import analysts because of the relative rigour given to nutrient feeding considerations.

Plans are one thing and actuality is another, even in so-called "controlled" centrally planned economies. Professor Schmidt discusses (Chapter 7) five-year production plans for the period 1981-85 relative to actual outcomes. That the essence of planning is crisis management as much as or more than long-term direction of the economy becomes especially evident in Chapter 3, as Brada analyzes how planners respond to harvest failures in terms of importing grain to sustain livestock herds and production.

Analysts of economic behavior in the centrally planned economies are wont to emphasize their institutional and ideological distinctions, but centrally planned economies, like other economies, are facing certain basic problems in economic survival. Scarcity requires these economies to confront problems of choice, though perhaps with a different style. In particular, predicting the import volumes of a country must be tied to the mechanisms used and the degree of success achieved in solving its food policy/economic growth dilemma. This holds particularly when a large share of resources of necessity must be allocated to feeding people. None of the centrally planned economies claim affluence, but China in particular is a poor economy. How successfully the country feeds itself has a direct impact on imports. Given its size, China's trade and food policies can affect the world terms of trade, but what happens to world prices as a result of China's participation in world agricultural markets can in turn affect China's ability to sustain imports. Data and information on China's economy are still too sparse to model econometrically the economy for purposes of test- ing hypotheses or prediction. For this reason, the article (Chapter 6) by Timmer and Jones heuristically discusses broad policy options available to the Chinese in an effort to analyze how China's role as a grain buyer may affect the world grain market.

DOING BUSINESS WITH CENTRALLY PLANNED ECONOMIES

Firms doing business with centrally planned econo- mies operate in a different environment than when they deal with private importers in market type economies. The state monopoly of foreign trade in principle vests unique power in the hands of importing agencies. More- over, these are public agencies operating within the framework of state socialism. In Chapter 8, Bunker, Jones, and Conley recognize this and point out that there are certain trading practices sometimes preferred by CPEs. Nevertheless, it is stressed that the same commercial principles that dictate satisfying the custo- mer in a free enterprise setting also apply when export firms deal with traders in the socialist countries. They seek to obtain as favorable a deal as possible in commercial terms. Several marketing strategies and business arrangements and practices are discussed in Chapter 8 in the practical context of what considera- tions apply in trade dealings with centrally planned buyers.

CONCLUSION

After reading this collection of papers on the cen- trally planned economies, the conclusion that will emerge is that agricultural trade relations with the centrally planned economies cannot easily be reduced to simple characterization. Unique circumstances among these countries is part of the reason, but their import behavior also will depend upon the policy response of the authorities that direct trade decisions. This policy response is difficult to predict because the import decision process involves multiple trade-offs and a complex negotiating process. Frequently, bureaucratic inertia will delay action and then once a decision is made, it may be in response to a crisis situation or so delayed that the magnitude of its impact is much greater than would be the case in a decentralized market economy setting. Indeed, if any single generalization can be made about the operation of a planned economy, it is not so much that central planning allows economic events to be anticipated or controlled, but more that decisions on economic matters are delayed until they have become a crisis. Ironically, controlled economies are frequently out of control and this principle reveals much about their agricultural trade behavior as well. This volume hopefully brings out a number of factors that underlie agricultural trade decision making in centrally planned economies. The contents fall far short of providing all the information needed to understand agricultural trade issues as they relate to the centrally planned economies but we hope we have at least made a step in that direction.

NOTES

1. See Josef Wilcynski. 1973, p. 208.
2. Income per capita, the projected rate of growth of GDP or GNP, the import reserves ratio, the share of imports in GNP and export fluctuations have been suggested as other considerations used by international lenders to assess the debt-service capacity of borrowing countries. (G. Feder and R.E. Just).
3. Another consideration is that central planning authorities have found that bilateral trade agreements are well suited to their purpose of enlisting trade into the service of the overall national economic planning process. Long-term arrangements reduce economic uncertainty which would otherwise complicate economic planning (Franklyn D. Holzman). Knowing in advance what will be available from foreign sources and what must be traded to secure these imports within a given stipulated timeframe reduces the uncertainty otherwise imposed on the planners by external world market developments. In mutual trade the planned economies resort to strict bilateralism, but in trade with the market economies the bilateral basis of exchange has been strongly favored also. These instruments are also useful as a means of ameliorating or side-stepping chronic shortages of convertible foreign currencies and on this basis the arrangements have played an extremely important role in trade with the developing nations who typically are also plagued with such difficulties.
4. In one sense bilateralism for political reasons may promote a higher level of trade than multilateralism. A political hindrance to free multilaterally conducted trade derives from the fact that domestic interest groups who are adversely affected by import competition concentrate and marshal their political influence in favor of trade hindering legislation. The diffused interests of consumer and export groups are less effective in carrying the argument for liberalization of trade. In bilateral barter type arrangements, however,

the necessity of trade being two way is more apparent and may make it easier for export workers to identify the necessity of imports.

5. A fourth issue is how diplomatic relations between Western exporters and the Socialist importers interrelates with commercial agricultural trade. See Roosa, Robert V., Michlya Matsukawa and Armin Gutkowski, East-West Trade at a Crossroads, New York, Trilateral Commission, 1982 for a discussion of this issue.

REFERENCES

C.I.A. National Foreign Assessment Center, Estimating Soviet and East European Hard Currency Debt. ER80-10327 (June 1980): 7, Table 4.

Crane, Keith. "Systematic Differences in Foreign Trade: A Key to Differences in the Cost of Polish and Hungarian Hard Currency Account Adjustment." Paper presented at American Association for the Advancement of Slavic Studies Annual Convention, Kansas City (October 1983).

Bornstein, Morris. "Economic Reform in Eastern Europe," in East European Economies: Post Helsinki. Joint Economic Committee, 95th Congress First Session (Aug. 25, 1977), pp. 102-135. esp. 131-132.

Brada Josef C. and Marvin R. Jackson, "The Organization of Foreign Trade Under Capitalism and Socialism," Journal of Comparative Economics. (June 1978): pp. 293-320.

Csikos-Nagy, Bela. The New Path of Hungarian Price Policy (Paper presented at AERI Seminar, Academy of Science of Hungary, April 1979), p. 1.

East-West Markets. "Eastern Approaches." (November 15, 1976): p. 4.

Feder, G. and R.E. Just."An Analysis of Credit Terms in the Eurodollar Market," European Economic Review, 9 (1977) pp. 221-243, esp. p. 235.

Foreign Agricultural Service, United States Department of Agriculture. Warsaw: Attache Report PL-0033 (August 1, 1980): p. 4.

Hardt, John P. "Long-Term Agreement: Some Considerations for Agricultural Trade." Paper prepared for the Conference on East-West Trade, Technology Transfer, and United States Export Control Policy, Institute of International Studies, University of South Carolina (March 1-3, 1983): 2.

Holzman, Franklyn D. International Trade Under Communism, New York: Basic Books Incorporated, 1976, p. 25.

Journal of Commerce. "Problems Mount in Eastern Europe." (December 28, 1981) pp. 3A, 9A.

Markish, Yuri. "Procurement Prices for Soviet
   Agriculture in the 1980s." in John L. Scherer,
   (ed.), USSR Facts and Figures Annual, Vol. 9, Gulf
   Breeze Florida: Academic International Press, 1985,
   pp. 216-223.
McCalla, Alex F. "Structural and Market Power Consider-
   ations in Imperfect Agricultural Markets." in Alex
   F. McCalla and Timothy E. Josling, editors. Imper-
   fect Markets in Agricultural Trade. Montclair, New
   Jersey: Allanheld, Osmun and Co., Publishers Inc.
   (1981).
_____, "Economic Problems Mount in Eastern Europe."
   Journal of Commerce. (December 28, 1981): 3A, 9A.
"Polish, Chinese Deals for Brazil." Seatrade (August
   1978).
Snell, Edwin M. "Eastern Europe's Trade and Payments
   with the Industrial West," Reorientation and Com-
   mercial Relations of the Economies of Eastern
   Europe. Joint Economic Committee, Congress of the
   United States (August 16, 1974): 685.
United Nations, (ECE), Appendix Table 1.C
Wilczynski, Joseph. Profit, Risk and Incentives Under
   Socialist Economic Planning. London; Macmillan
   Press, Ltd. (1973): Chapter 9.
The Wall Street Journal. (February 11, 1982): p. 28.

2

The Effect of the U.S.-Soviet Bilateral Trade Agreement on the World Market: Implications for United States Policy

Alan J. Webb and Bob F. Jones

INTRODUCTION

The variability of world prices and growing concern among importing countries about the availability of grain supplies on the world market in the past ten years has resulted in a more frequent use of bilateral trade agreements. The current specification of these agreements in the grain trade generally provides for the flow of one or more agricultural commodities between two countries for a given time period. Although these agreements usually do not carry the weight of formal treaties, they guarantee supplies to concerned importing countries while assuring exporters markets for their grain.

Trade agreements have been especially important in the grain trade with the centrally planned economies. Roughly 80 percent of all wheat committed through bilateral agreements in 1980 was between the four major wheat exporters (Argentina, Australia, Canada and the United States) and the centrally planned economies (principally China, Poland and the Soviet Union). Even the United States, which has long resisted the use of bilateral agreements, has sought to use long-term agreements with centrally planned countries.

This chapter will analyze the role played by bilateral trade agreements in world grain trade while focusing on their importance to centrally planned economies. This analysis will be followed with an examination of the market conditions leading up to the 1975 U.S.-Soviet agreement, the value of the agreement to both participants, and the effect of the agreement on world grain trade. Emphasis is on the first U.S.-Soviet five-year agreement and its two one-year extensions ending in

---

*The views are those of the authors and do not necessarily reflect those of the U.S. Department of Agriculture.

1983. The second five-year agreement which took effect in September 1983 has changed the quantities and the mix of grains but has not significantly altered the key elements of the first agreement. The final section will discuss the value of additional U.S. trade agreements.

Structure and Objectives of Bilateral Trade Agreements

Bilateral grain agreements specify the flow of one or more commodities between two countries for a given time period. These agreements are different from those negotiated prior to World War II under the auspices of the Reciprocal Trade Agreements Act of 1934. The latter were reciprocal agreements for the reduction of tariffs on specific commodities important in the trade between two countries. The current use of bilateral agreements in world agricultural trade focuses on the guarantee of a minimum level of commodity trade between two countries.

There are four important elements to be specified in terms of a bilateral commodity agreement; duration, quantity, price and degree to which the parties are bound by the agreement. The objectives of the countries signing the agreement will determine the terms.

Duration. Agreements currently average three to five years in duration. Countries which are concerned about guaranteeing long-term supplies or markets will favor longer-term arrangements.

Quantity. Most agreements are flexible specifying a minimum annual purchase. There may be an upper limit also, as in the U.S.-Soviet agreement, which can be exceeded if it is mutually agreed by both parties. Some agreements set minimum purchase requirements over the length of the agreement as well as annual minimums.

Prices. Specification of a price is generally left to be determined at the time of the actual sale. However, the problem of inconvertible currencies has led the centrally planned economies to negotiate bilateral barter arrangements for many of their mutual commercial transactions. A price is implied in these barter agreements since the quantity of one commodity is specified in terms of another.

Degree of binding. The degree to which countries are bound by the terms of an agreement varies with the type of agreement. There are a number of unwritten "understandings" between governments or trading agency officials for the exchange of information on import intentions and export availabilities with, perhaps, a commitment to exchange a given quantity which could be considered bilateral agreements. The U.S. government has had multi-year trade understandings for export of U.S. agricultural commodities with Japan, Poland, Israel, Norway and Taiwan.[1]

For most purposes, however, bilateral agreements are written commitments with a clearly specified quantity and time period. Because the grain of most countries is traded by government or quasi-government agencies, the written agreements carry government commitments to complete the transactions. The U.S. government, in contrast, is not directly involved in the delivery of grain to foreign markets. Most U.S. government bilateral agreements are "entitlements" which guarantee an importer access to U.S. grain markets as well as priority in obtaining the quantity specified in the agreements. Although there is no supply guarantee, the U.S. is expected to do all that is reasonable within its power to make the agreed quantity available.

The Effect of Bilateral Trade Agreements on Price and Trade Variability

The primary difference between a bilateral trade agreement and other exchanges on the world market is that the former is a government-to-government arrangement which establishes a level of trade over a particular time period. It is possible that the series of exchanges which take place under a bilateral agreement would have taken place without the agreements. In this case, the bilateral agreement has merely served to give government sanction to a trade flow which would have occurred anyway.

Trade agreements, however, may distort existing trade flows without affecting the level of total imports or exports significantly. Transportation and handling costs may increase somewhat if the bilateral trade pattern is less optimal than the preceeding pattern.

Bilateral trade agreements, in general, will not alter the underlying supply and demand schedules from which import demand and export supply functions of individual countries are generated. Countries desiring to negotiate bilateral agreements attempt to anticipate their future import needs or export supplies before agreeing to a long-term contract. Variability of production due to weather and other factors and the associated volatility of world prices make it to a country's advantage to negotiate agreements totalling less than the country's anticipated future trade. The remainder can then be imported or exported on the open market.

The effect of bilateral agreements on world price variability will therefore depend on how these agreements affect the variability in an individual country's trade as well as their impact on the price responsiveness of that trade. A graphical exposition will help illustrate these effects. Figures 2.1 and 2.2 show the effect of annual changes in excess supply or demand on equilibrium prices in a multicountry world market in

32

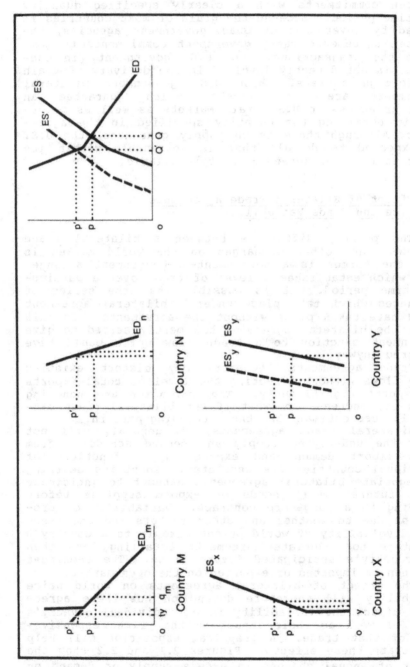

Figure 2.1 Effect of a decrease in foreign excess supply on an importing country with a bilateral trade agreement.

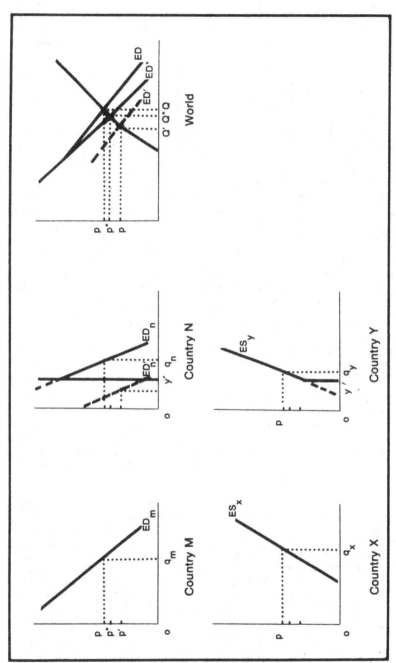

Figure 2.2 Effect of a decrease in domestic excess demand on an importing country with a bilateral trade agreement.

which a bilateral trade agreement inhibits market adjustment in an importing country. Figure 2.1 shows the effect of a decrease in foreign export supplies and Figure 2.2 shows the impact of an increase in domestic supplies (and therefore a decrease in import demand) within the contracting importing country.[2] The two scenarios yield different results and thereby help clarify some of the issues which have arisen in discussions of the desirability of establishing bilateral trade agreements.

The excess supply and excess demand schedules for four countries are shown in Figure 2.1. Countries M and N are importers and countries X and Y are exporters. Countries X and M have signed a bilateral trade agreement to exchange a quantity of oy per year over a number of years. Consequently, the excess demand and excess supply schedules of M and X respectively become perfectly price inelastic when the quantity traded falls to the level of the agreement. In the world market, excess demand (ED) and excess supply (ES) schedules are aggregations of the individual country curves and therefore become less price elastic at the points at which the schedules for countries M and X become constrained by their bilateral agreement. Note that the bilateral agreement has not affected the level of the equilibrium price, P, or the amounts exchanged in the world or country markets.

Suppose that there is a major crop shortfall in exporting country Y resulting in a shift in Y's excess supply from ES_y to ES_y'. World excess supply reflects this crop shortfall as it shifts from ES to ES'. World price increases from P to P' and the quantity traded falls from OQ to OQ'. At the new higher price, P', country M finds itself overcommitted because it would have chosen to import only quantity ot had it not been required to purchase oy under the terms of its trade agreement. Country M therefore purchases unwanted quantities which raises world prices higher than they would have been without the agreement.

Figure 2.2 shows the same four country world markets with essentially the same equilibrium conditions except that countries Y and N now have a bilateral agreement for the exchange of quantity oy'. This time, assume that country N has a large domestic crop which, under normal circumstances, would reduce its excess demand from ED_n to ED_n'. World excess demand would shift by a similar amount from ED to ED' causing the equilibrium price and quantity traded to fall to P' and Q' respectively. The terms of the bilateral trade agreement, however, prevent country N from reducing its quantity of excess demand below oy'. This restriction results in a shift in aggregate excess demand of only ED to ED* and a corresponding smaller fall in world prices and quantities traded (to P* and Q* respectively). In this case, the bilateral agreement has reduced the adjustment in world prices required for a given shift in excess demand although substantial internal price adjustments might be required by country N to absorb the additional quantities.

The effect of an overcommitment to bilateral agreements on price variability is inconclusive since it depends on the source of the change in the world market which results in the overcommitment. The effect of a bilateral agreement in both cases is to maintain trade at a minimum level. In Figure 2.1 it prevents the full transmission of world market price changes onto the domestic market because it prevents country M from making a complete adjustment to the short world supply situation. The effect is reversed in Figure 2.2. The trade agreement prevents the full transmission of the price effects of a large domestic crop onto the world market.

An overcommitment by an individual country will result, in any case, in a divergence between domestic and world prices. In importing countries, the effect of an overcommitment is to lower the domestic price relative to the world price; in an exporting country an overcommitment will have the opposite effect, i.e., it will raise the domestic price relative to the world price.

A country which seeks to avoid the price distortions which accompany an overcommitment must establish a minimum total level of bilateral trade commitments which is well below future expected trade. The problem from the viewpoint of an importing country (as illustrated in Figure 2.3) is to determine the lowest probable level of its expected annual imports to be used as the upper bound on its bilateral trade commitments. There are two sources of uncertainty in determining expected future trade levels; 1) the potential variation in domestic supply and utilization, and 2) the potential variation in the rest of the world's supply and utilization. Importing country policymakers in Figure 2.3 need to know the joint probability distribution of their excess demand schedule (ED) and the excess supply schedule facing them (ES). Assuming that their two schedules are independently distributed, the joint probability distribution is the product of the individual probability distributions. Let policymakers set a lower bound on expected excess demand (ED_L in Figure 2.3) such that the probability of a shift in excess demand to the left of ED_L is 30 percent. Given a similar 30 percent lower bound on expected excess supply (ES_L), the probability that both will fall below their respective 30 percent lower bounds is 9 percent. The probability that imports will fall below mean level imports (M^*) is depicted by the cumulative probability distribution in the lower portion of Figure 2.3. The intersection of

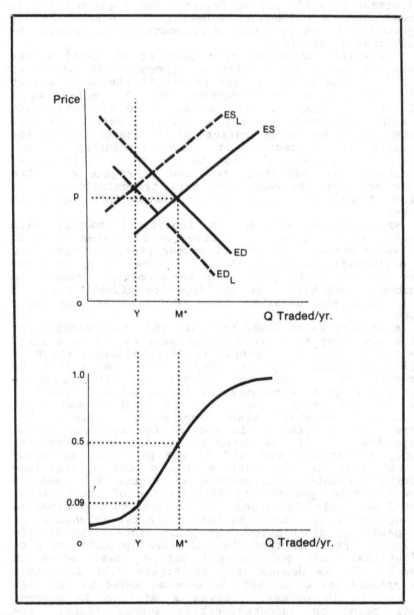

Figure 2.3 Assessing the probability of an overcommitment
to bilateral agreements in an importing country.

the lower bounds (ES_L and ED_L) correspond to a probability of 9 percent in the bottom panel of the figure.
The risk of an overcommitment to bilateral agreements
can be limited to 9 percent if combined commitments are
restricted to no more than quantity oy.

The proportion of a country's trade which can be
"safely" committed to bilateral agreements will depend
upon the excess demand and excess supply elasticities as
well as the probability distributions of the two functions. A small importing country with a stable inelastic excess demand schedule, for example, will be able to
commit a much larger proportion of its imports to bilateral agreements at the same level of risk than will a
large importer with a highly variable and price elastic
excess demand function.

Policy Alternatives Available to
Overcommitted Grain Traders

Overcommitment of bilateral agreements can easily be
avoided given a rudimentary knowledge of the trade parameters facing individual countries. For this reason,
bilateral agreements have not been viewed as a major
constraint on world trade. Even in the few cases in
which an overcommitment might have occurred, there are
ways of avoiding or mitigating price and trade distortions which might result.

Three broad policy options are available to decision
makers in an overcommitted country. The country can; 1)
abide strictly by the terms of its international commitments and adjust its domestic market to compensate, 2)
it can violate the intent (but not the letter of the
agreement) by transshipping the amount of shortfall
coverage, or 3) it can renege on one or more of its
commitments. For most market economies with a direct
link between domestic and world markets, arbitrage will
prevent a wide deviation in the margin between domestic
and world prices. The first option is viable, therefore, only to the extent that the government is willing
to adjust its holdings of stocks (if it has any) to meet
its bilateral commitments. This action may be sufficient as long as trade commitments are only slightly
greater than desired trade, but for a large overcommitment, transshipment is often the least costly solution
from a government viewpoint. Transshipments avoid
possible international repercussions of a broken agreement (i.e., it might be difficult to secure future
agreements) and they are less costly than large stock
accumulations.

The action taken also will depend on the structure
of the grain marketing system. Independent trading
companies, for example, may have little incentive to
abide by an agreement to which they are not a party and

for which their individual prestige and future sales are not at stake, particularly if transshipments will involve substantial outlays on their part. (This, no doubt, helps account for the apprehension with which many U.S. grain trading firms viewed the proliferation of trade agreements signed by the United States in the late 1970s.) The action taken to adjust for an over-commitment will, therefore, depend on whether the government or its trading agency can enforce compliance to a long-term agreement. Countries with grain boards (Argentina, Australia, Canada, South Africa) or food import agencies (Japan, Taiwan, Mexico) can enforce long-term agreements and pass any costs associated with an over-commitment along to producer members, consumers or to taxpayers. The legal responsibilities are less clearly defined in countries such as the United States where government agreements must be carried out by inde-pendent private traders.

The institutional framework for trade in centrally planned economies is different from most developed mar-ket economies. Rigid internal prices are not linked with the world market and only broadly reflect the availability of products relative to demand in domestic markets. These countries participate in world agricul-tural trade through state trading agencies. Trade in the short run, therefore, is primarily a function of policy decisions (e.g. imports may depend on planning targets for the level and mix of livestock feeding). There tends to be relatively little trade response to a change in world prices (note the price elasticities quoted in studies by Seeley, Tyres, and Longmire and Dunmore). Inelastic excess demand or supply means that centrally planned economies are not likely to become overcommitted due to fluctuations in prices on the world market (as shown in Figure 2.1) but may become over-committed because of shifts in domestic supplies or demands (Figure 2.2). These fluctuations may arise from policy changes (or miscalculations) as well as from changing weather conditions.

The importance of the 1975 U.S.-Soviet Grains Agree-ment deserves careful consideration not only for deter-mining the value, if any, of the agreement itself but also as a standard against which future agreements, including the second U.S.-Soviet five-year agreement signed in 1983, might be measured. The analysis of the usefulness of the agreement will also provide valuable insights as to the desirability and structure of other agreements.

A large bilateral trade agreement (or a series of agreements for a large quantity) with a large centrally planned trader such as the Soviet Union, could stabilize world prices if; 1) as shown in Figure 2.2, trade agree-ments prevent domestic fluctuations in availabilities from being fully translated into changes in transactions on world markets, 2) the size of the agreement precludes the use of the transshipments as a means of evading the intent of the contract since alternative markets or sources would be more difficult to arrange for large quantities, and 3) the agreement itself has the effect of ensuring that decision-makers within the centrally planned economy itself would generally adhere to the policies which initially gave rise to the agreement.

No bilateral grains agreement has had a larger impact on world grain markets than the five-year 1976-1981 U.S.-Soviet accord and its two one-year extensions signed in 1975. The agreement was capable of meeting all three of the functions enumerated above and therefore by itself could have been instrumental in bringing a degree of calm back to a nervous and unstable grains market.

The following sections will examine the atmosphere of grains markets in the early 1970s and the events leading up to the signing of the five-year U.S.-Soviet grains agreement. The objectives of the agreement and its structure will be analyzed and conclusions will be drawn as to whether the U.S.-Soviet agreement did or did not play a major role in stabilizing grain markets. Finally, the desirability and possible structure of future agreements with the Soviet Union and other centrally planned economies will be considered in light of evolving world grain markets and the changing objectives of the major participants.

The World Grain Trading Environment of the Early 1970s

In the ten years prior to 1972, U.S. agricultural exports increased from five billion dollars to the eight billion dollar level. Throughout the 1950s and 1960s growth in exports and domestic consumption lagged behind increases in grain production. As a result U.S. grain stocks continued to grow. Carryover stocks of U.S. wheat and feed grains reached a peak of 94.9 mmt in 1962 when they represented about 55 percent of annual utilization (domestic consumption plus exports). A large proportion of U.S. grain stocks was owned by the Commodity Credit Corporation (CCC) which acquired grain as a result of government price support operations.

U.S. grain stocks declined temporarily during the world food scare of 1965-67 -- a time when the Soviet Union and India made large grain purchases on world markets to cover production shortfalls. By late 1968, however, world grain production had recovered, (particularly in the Soviet Union which dropped out of the world market) and U.S. stocks again began to accumulate.

With the existence of large grain stocks under CCC ownership, the government was faced with three alterna-

tives; 1) use more effective production restrictions, 2)
continue to accumulate stocks which required large outlays for interest and storage payments, or 3) push subsidized exports more aggressively. Since farmers were
unwilling to accept tougher production controls and
continued accumulation of stocks appeared to be a very
expensive (as well as only a temporary solution) the
decision was made to increase efforts to export grain.
Funding for noncommercial exports under P.L. 480
increased in the early 1970s and export subsidy payments
increased from $679 million in 1969 to $2,750 million in
1973 (Cochrane and Ryan 1976).

Twenty years of experience with overproduction, low
grain prices, large grain stocks under government control, and problems in expanding commercial exports had
not conditioned policymakers to anticipate the rapid
expansion in export demand that occurred in the early
1970s. A number of circumstances and structural changes
combined to make the 1972 Soviet grain purchase the most
disruptive in the recent history of world grain trade.

The purchase itself was presaged by a change in
Soviet policy. In the five-year plan adopted in the
Soviet Union for 1971-75, decisions had been made to
increase meat and poultry production by a significant
amount. But in the mid-seventies, the decision was made
to import large amounts of grain to meet production
shortfalls. This decision represented a dramatic change
in policy as typically livestock herds had been reduced
when grain supplies fell well below needs.

In the summer of 1972 when it became obvious that
Soviet grain production would be insufficient to maintain grain consumption levels, Soviet buyers came to the
United States, seeking grain in large quantity. Initially, it was not known how much they wanted to buy
but, in any case, they found willing sellers. Commercial firms were eager to accomodate them and the government was willing to facilitate the sale as this was an
opportunity for it to get out of the grain storage business. Furthermore, the United States had sold 1.7
million tons of wheat to the Soviet Union in 1963 and
Canada had sold much larger annual quantities in the
period from 1963 to 1966 without causing any serious
disruption to the market.

Yet a number of other events contributed to the
rapid transformation of the world grain market from one
of abundance to one with extremely tight supplies. On
the supply side, the decline in fish meal production in
Peru and reduced grain and sunflower production in the
Soviet Union brought sharp reductions in grain and protein meal availability. In addition, burgeoning stocks
had caused two of the world's major grain exporters
(Australia and Canada) to institute drastic one year
measures in 1970 to reduce wheat production and reduce
stocks. The United States, of course, had had policies

in place which held down acreage planted for many years. Hence, policy measures and weather combined to reduce the supplies which would have been available for domestic use or export in a few key countries.

On the demand side, steady growth in real income and population in many countries gave underlying strength to the market. Also, policies had been instituted by both importers and exporters to limit the effect of a change in world grain prices on domestic markets. The formation of the European Community and the creation of the Common Agricultural Policy was a development which reduced the ability of a major portion of the market to respond to a rise in world prices. Japanese trade restrictions on wheat and rice as well as subsidies on grain consumption in a large number of developing countries also were significant in preventing market adjustments.

A number of factors also delayed or diminished the response of the U.S. grain marketing system to the Soviet purchase. The 1972 sale was the largest grain sale in history to that time, eventually reaching over 11 mmt of grain, most of which was wheat. Yet the full dimension of the sale was not known until much of the grain had been sold. Certainly, the closed Soviet economy and the lack of informaton concerning purchase intentions within the Soviet Union purchasing agency were a major source of uncertainty. This, coupled with U.S. willingness to sell at an essentially fixed price (through the use of export subsidies) seemed to give the bargaining advantage to the Soviet Union. Grain companies were willing to contract the sale of grain which they did not yet own (a normal procedure) because they had access to the futures markets in which they could hedge sales but, more importantly, they knew the CCC held large grain stocks and they had the commitment of the government that prices would be assured through the export subsidy system. Competing U.S. export companies did not know the level of sales made by their competitors because timely and complete information about the amount of grain being sold to the Soviet Union and other buyers was not available. The transmission of the purchases into an upward movement in futures prices was inhibited by the subsidy on wheat exports. The crucial informational elements were missing for the efficient operation of a competitive grain marketing system.

The United States undertook measures, of which the U.S.-Soviet agreement was one, to rectify this situation. When it became clear, in the summer of 1972, as to how much grain had already been sold to the Soviet Union, the export subsidy guarantee was ended and has not been used since.

Efforts were also directed at the export reporting system. Under the system which prevailed during the 1950s and 1960s grain exporters were required to file

reports of actual shipments of grain with U.S. customs.
This reporting system gave the government no control
over the volume and terms of sale for export. The
information became public a month or longer after
shipments had been made. No reporting of sales for
future delivery was required nor was it considered
necessary when exports were relatively small and stocks
were large.

That system, however, was inadequate for the export
situation that developed in 1972. The general tightening of supplies and uncertainty about the amount of
grain being contracted for export led to modifications
in the export reporting system starting June 13, 1973
(ERS-USDA November, 1974). Exporters were required to
make weekly reports of outstanding sales. Two weeks
later an embargo was placed on new sales of soybeans,
soybean meal and cottonseed. Shortly thereafter the
embargo was replaced by a licensing system whereby new
sales of the embargoed products required Commerce
Department approval. The licensing program remained in
effect until October 1, 1973.

Unfavorable reactions from many foreign customers
emphasized the seriousness of embargo and license cut-
backs. Responsibility for monitoring export sales of
agricultural commodities was moved to the Department of
Agriculture.

Weekly reporting of export bookings appeared to be
sufficient control over exports during the early part of
1974. However, as U.S. crop conditions continued to
deteriorate during the summer of 1974 and export orders
increased, the USDA required daily reports effective
September 12, 1974. On October fourth it was announced
that sales of corn and wheat to the Soviet Union were
suspended until discussion could be held between the two
governments.

On October 7, 1974 the USDA announced a "voluntary
cooperation system" for reporting prospective export
sales. This system required daily reports and prior
approval of sales beyond a specific size to a single
buyer. When cumulative sales to one buyer during a week
reached double that figure, sales were to be reported
and prior approval sought. Two weeks later sellers were
advised to make sales "subject to FAS approval" and to
inform their buyers of this new requirement.

This system of reporting and "prior approval" is
currently in effect, although it was suspended in March
of 1975. It was reinstated in July 1975 as uncertainty
about export sales reemerged when information about 1975
Soviet crop prospects developed. The system has undergone only minor modifications since July 1975.

The elimination of export subsidies and the establishment of export reporting procedures were directed at
correcting problems in the U.S. marketing system but
they did not address the source of the disinformation --

the Soviet Union. The five-year U.S.-Soviet long term agreement was intended, from the U.S. perspective, to induce greater Soviet cooperation in stabilizing world grain markets.

Events in the world grain market in 1975 precipitated the signing of the five-year grain agreement. In particular, the Soviet grain crop fell to 140 mmt, or 75 mmt below the planned production level. During the second half of July 1975 the Soviet Union bought 9.8 million tons of grain from the United States. World grain prices began to rise rapidly. On August 11 grain companies were asked to voluntarily withhold further sales of grain to the Soviet Union until fall crop prospects were more certain. On September 7, 1975 sales to Poland were also included in the moratorium. It was also announced that the moratorium would remain in effect until the October crop report. The moratorium was not lifted until the signing of the five-year agreement.

The United States and the Soviet Union entered into the five-year bilateral grain agreement in October of 1975. It represented a departure from traditional U.S. trade policy which had relied heavily on multilateral forums to provide broad guidelines for conducting trade with other nations.

Purposes and Structure of the 1975
U.S.-Soviet Grain Agreement

The agreement specified that the Soviet Union would purchase in each twelve month period beginning October 1, 1976, 6 mmt of wheat and corn in approximately equal proportions. An additional 2 mmt each year could be purchased without consultation unless the U.S. government determined that the United States had a supply of less than 225 mmt. The United States could authorize sale of larger amounts and the Soviet Union could request the purchase of more wheat and corn. The Soviet Union agreed to space their purchase and shipments as evenly as possible over each twelve month period. It was agreed that the parties would hold consultations concerning the implementation of the agreement at six month intervals. These consultations would lead to periodic inspection of Russian crop conditions by U.S. observers and the exchange of other information about Soviet needs for grain.

The official U.S. explanation for the signing of the agreement implies three major U.S. objectives (White House Press Secretary, 1975). First, the United States wanted to stabilize the flow of grain from the United States to the Soviet Union. It was hoped that the minimum purchase requirement would induce the Soviet Union to carry larger reserve stocks of grain (see

Hilker et.al., 1981). Also, more even spacing over the crop year of Soviet grain purchases would help relieve seasonal pressure on U.S. grain markets. Secondly, the United States sought to make Soviet purchases more predictable even if stability could not be attained. Consultations and periodic inspections of the Soviet Union's grain crop during the growing season were intended to reduce some of the price fluctuations generated by uncertainty and misinformation. Finally, the United States wanted to secure a market for U.S. wheat as well as coarse grains by requiring that the Soviet Union purchase at least 6 mmt of wheat. Without this requirement, the Soviet Union might have purchased primarily coarse grains from the United States and taken only a limited quantity of U.S. wheat imports. The United States, which holds a 70 percent share of the world market for corn, has very little competition from other suppliers; but, the United States has only a 35 to 40 percent share of the world wheat market in which Canada, Austrialia, Argentina and the European Community are major competitors.

Performance of the Agreement

The 1975 Long Term Agreement (LTA) was to provide additional assurance that the events of 1972 would not be repeated. Some analysts have maintained that the agreement was unnecessary window dressing and that the most important measures had already been taken, (i.e. the elimination of export subsidies and the establishment of an export reporting procedure). Yet the earlier measures, though correcting serious defects in the U.S. grain marketing system, did not attempt to address the source from which the shock to the system had originated. The LTA did, at the very least, provide a framework for greater cooperation between the two countries and there was evidence to suggest that the agreement, if correctly specified, might induce a change in the trading and stockholding policies of the Soviet Union. The foregoing graphical analyses suggests that a trade agreement could induce such a change; the following simulation of the Soviet Union's feed-livestock economy during the late seventies provides further justification for the use of a trade agreement with the Soviet Union.

A Simulation Analysis of the
Five-Year Agreement

A relatively simple econometric model was developed to estimate annual grain import needs for the Soviet Union and determine th effectiveness of the 1975 LTA relative to other scenarios (Hilker et al. 1981). The aggregate demand function for grain made consumption in the Soviet Union a function of grain used for human consumption, grain used for seed, grain used for industrial uses, grain wasted and quantity of grain fed to livestock and poultry. Grain used for feed was a function of hog and chicken numbers. Since only about 15 percent of grain consumption is by cattle in the Soviet Union, cattle were not included in the consumption equation. Because of inadequacies in the data at the time the model was constructed, the feeding rate per animal unit was assumed to remain constant over time. Hog numbers were made a function of grain production to year t and hog numbers in year t-1 under the assumption that hog numbers are adjusted to grain availability with a lag. Likewise, chicken numbers were a function of grain production and the number of chickens the previous year.

The aggregate supply function for grain considered production a function of area harvested and yield. Yield was a function of trend and a random variable which was assumed to capture the effects of weather on yields.

Price was not included in any of the equations primarily because of a lack of good information. Although there is some evidence the Soviets do respond to changes in the price of grain, this is not considered a serious limitation of the model, since price plays a much smaller role in an economy where demand and supply decisions are made by plan with internal prices fixed by the government.

The quantity of grain imported by the Soviet Union was a function of grain consumption, grain production and changes in grain stocks. Stocks were assumed to be zero in the inital year because of lack of better information and the fact that the 1975 crop was so small that stocks were probably reduced to pipeline levels.

A probability distribution of annual grain imports was estimated using the grain import simulator. In order to study the affects of the agreements, various scenarios were developed to represent different decision rules which might be followed by the Soviet Union. A similar model was developed for Eastern Europe using the same type of supply, demand and import functions in order to account for principal changes in trade flows, because of the close historical trading relationship between the Soviet Union and Eastern Europe. One assumption in the link between the two models is that the Soviet Union is the first supplier to Eastern Europe, (i.e. in years when the Soviet Union has excess grain it would be shipped to Eastern Europe with Eastern Europe buying less from other outside sources).

Was it reasonable to expect that the 6 mmt minimum purchase would narrow the distribution in year-to-year Soviet Union grain imports? Three different scenarios provide insights. In the first, the Soviet Union and

Eastern Europe import grain with no grain import commitments from the United States. The second assumes the Soviet Union has a 6 mmt agreement with the United States such as the 1975 LTA. The third scenario assumes the Soviet Union has 9 mmt of commitments and Eastern Europe has 4.5 mmt of commitments. A 9 mmt option for the Soviet Union was considered because it was noted that the Soviet Union has not imported more than about 65 percent of their needs from the United States in any one year. Nine mmt of total imports would allow 66 percent of Soviet import needs for the late seventies to be purchased from the United States. The 4.5 mmt minimum amount that Eastern Europe must import in the last scenario represents the minimum amounts that Eastern European countries would have to import to fulfill agreements and "understandings" in existence in 1976.

The results show that the U.S.-Soviet grain agreement alone would not significantly increase world exports of grain or reduce variability of shipments. When only the Soviet Union is committed to buy U.S. grain, the average combined imports of the Soviet Union and Eastern Europe increased from 17 mmt per year to 18 mmt per year and the variation of grain imports does not change significantly. However, when Eastern Europe is also constrained by grain purchase agreements with the rest of the world including the United States, as in the third scenario, average combined grain imports increased from 17 mmt per year to 19.8 mmt per year and the range of probable grain imports was reduced by 40 percent. Therefore the model suggests that grain agreements with Eastern Europe which maintain the level of their purchases from non-Soviet sources when the Soviet Union has a large exportable surplus is a necessary condition for reducing variability in U.S.-Soviet grain trade.[3]

These results indicate the combined agreement requirements in scenario 3 would have the effect of causing the Soviet Union to carry larger stocks of grain and that this would have a stabilizing effect on world grain trade. The simulation results show mean grain stocks of 1.44 mmt when no commitments are in effect. When the U.S.-Soviet agreement and others are in force for both the Soviet Union and Eastern Europe, mean grain stocks increase to 7.82 mmt. However, the standard deviation of Soviet stocks increases as the mean level increases, suggesting, as expected, that more of the adjustment to changing Soviet production occurs through adjusting the size of domestic stocks rather than through adjusting import purchases.

The third scenario of the simulation is the most accurate reflection of Soviet and Eastern European imports in the late 1970s. Table 2.1 shows that even this scenario, which was intended to portray a reasonably high import situation, tended to underestimate Soviet and Eastern European import needs as four consec-

TABLE 2.1
Soviet and Eastern European Total Grain Imports, 1971-1984.

Year	Soviet Union				Eastern Europe			
	Imports		Exports	Net Imports	Imports		Exports	Net Imports
	Total	From U.S.			Total	From U.S.		
1971	8.1	2.9	6.9	1.2	10.4	1.3	2.3	8.1
1972	22.7	13.7	1.8	20.9	9.7	2.8	3.7	6.0
1973	11.2	7.9	6.1	5.1	10.6	2.1	4.7	5.9
1974	5.5	2.3	5.1	.4	10.1	2.9	2.8	7.3
1975	26.0	13.9	.6	25.4	13.2	5.8	4.1	9.1
1976	10.8	7.6	3.1	7.7	14.9	6.8	4.0	10.9
1977	18.8	12.5	2.1	16.7	13.5	5.1	3.7	9.8
1978	15.7	11.1	2.8	12.9	16.4	5.7	3.2	13.2
1979	31.2	9.8	.7	30.5	17.1	11.6	3.4	13.7
1980	35.3	8.7	.6	34.7	18.1	8.7	4.0	14.1
1981	45.9	14.5	.7	45.2	14.0	4.3	4.4	9.7
1982	31.9	6.2	.6	31.3	9.0	1.8	5.8	3.2
1983	32.5	13.8	.6	31.9	8.0	1.5	5.4	2.6
1984[a]	51.4	21.8	1.1	50.3	7.2	1.1	7.1	.1

Source: United States Department of Agriculture.

[a] Preliminary

utive Soviet crop shortfalls beginning in 1979 led to
growing Soviet grain imports. The minimum export
requirement (6 mmt) of the agreement, therefore, appears
to be largely irrelevant to the trade which took place.
The upper level of the agreement (8 mmt) became relevant
only because the United States suspended grains sales to
the Soviet Union in excess of this amount from January
1980 to April 1981. The sales suspension appears to
have had little effect on net Soviet grain imports
although Soviet imports from the United States were
probably well below what they would have been without
the embargo. Because Soviet imports for the late seventies and early eighties were substantially above the
levels specified in the 1975 agreement, it is difficult
to determine whether the agreement had any effect on
Soviet grain management practices. This does not necessarily mean that there were no changes in Soviet practices. As the subsequent sections point out, the Soviet
Union may have moved toward greater import dependence
had not Soviet crop failures and the U.S. sales suspension revealed the vulnerability of the Soviet Union to
disruptions in grain supplies.

Changes in Soviet Grain Management Practices

Soviet grain management practices have undergone
major revisions in the past fifteen years as the Soviet
Union has sought to raise per capita consumption of meat
and livestock products. A simple way of determining
whether any major changes took place is to examine the
variability of Soviet production, utilization stocks and
imports before and after the implementation of the first
Long Term Agreement. If the agreement was effective in
inducing the Soviet Union to transmit less variability
into world markets, the standard deviaton for Soviet
imports should decline in the second period while one or
more of the standard deviations of the other balance
sheet components should increase. Given that Soviet
production variability is largely weather related and
that the policy objective of the Soviets since 1970 has
been to reduce the adjustments in the livestock sector
which have resulted from variable feed availabilities,
the major adjustments in the domestic Soviet grain economy would then have to take place in stocks.
Table 2.2 presents the standard deviations, means
and coefficients of variation for Soviet balance sheet
components for wheat, coarse grains and total grains for
the five-year period preceding the agreement and the
subsequent seven-year period when the first U.S.-Soviet
LTA was in place.[4] The standard deviations for total
grains show no significant changes between the two
periods but a few of the balance sheet components for
wheat and coarse grains change significantly. The

TABLE 2.2
Changes in the Variability of Soviet Balance Sheet Components
After the 1976 U.S.-USSR LTA

	Standard Deviation		Mean		Coefficient of Variation	
	I	II	I	II	I	II
Coarse Grains						
Production	16.7	15.0	82.3	91.5	.20	.16
Total Util.	13.2	8.2	89.1	105.3	.15	.08
for Feed	8.5	5.1	59.5	71.4	.14	.07
Net imports	5.8	7.3	6.3	13.8	.92	.53
from U.S.b	3.3	3.0	4.7	6.7	.70	.44
Beg. Stocksb	.9	2.1a	6.6	3.8	.14	.55
Wheat						
Production	16.5	12.6	88.9	93.7	.19	.13
Total Util.	4.3	8.6a	93.7	104.6	.05	.08
for Feed	4.7	8.0a	34.4	43.2	.14	.19
Net imports	7.5	7.1a	3.9	11.3	1.93	.63
from U.S.b	3.7	1.4a	3.4	3.9	1.09	.36
Beg. Stocksb	6.9	6.0	12.6	5.9	.55	1.02
Total Grains						
Production	29.2	23.9	171.3	185.2	.17	.13
Total Util.	14.4	8.8	182.8	210.2	.08	.04
for Feed	6.8	6.3	93.9	114.6	.07	.05
Net imports	11.7	13.3	10.2	25.1	1.15	.53
from U.S.b	5.6	3.6	8.1	10.6	.69	.34
Beg. Stocks	7.4	7.4	19.2	10.3	.39	.72

I - Crop years 1971/72-1975/76. II - Crop years 1976/77-1982-83.
aSignificantly different from the 1971/72-1975/76 period level (10% level of confidence).
bSoviet grain stocks data are very unreliable because they are based on information and estimates of other Soviet balance sheet components. Errors in the other components will therefore be transmitted to and magnified in the stock estimate.

increase in the standard deviation for coarse grains stocks[5] and the fall in the variability of Soviet wheat imports from the United States are changes in the desired direction, from the U.S. perspective, but there are not enough significant changes in the variability of other components to infer a change in Soviet grain management practices. The reasons for the increase in the variability of Soviet wheat utilization are not apparent.

One reason the effects of a possible change in Soviet grain management practices may not show up in the standard deviations is the change in the mean levels of the balance sheet components between the two periods. This is important because the Soviet Union suffered a series of crop failures beginning in 1979 which inhibited their potential for shifting a portion of their variability in imports to stock changes. The coefficient of variation, which gives the standard deviation as a proportion of the mean, provides an indication of how variability in different balance sheet components has changed relative to the average size of those components. The coefficients of variation in the last two columns in Table 2.2 show the changes which are consistent with what would be expected if the Soviets had decided to reduce the variability in their imports. For both wheat and coarse grains, the coefficients of variation for stocks increase sharply following the implementation of the agreement while the corresponding coefficients for imports fall. From the standpoint of the world grain market, a seven to eight million ton variation in Soviet purchases is far less disruptive on a volume of 226 mmt (in 1984) than it was twelve years earlier on a volume of only 137 mmt.

The basic objective sought by the United States in the LTA, i.e. a reduction of the variability of Soviet grain purchases, was at least achieved in a relative sense. Whether this was in whole or in part the result of the agreement is another question. Many of the changes in Soviet grain management practices were the result of policy decisions and internal constraints which may have led to a similar result despite the agreement. The 1980 U.S. grains embargo of the Soviet Union is a separate but related event which looms large in its impact on Soviet grain policy. The embargo revealed the vulnerability of Soviet grain supplies as well as the value of long-term agreements and should have provided a strong impetus for larger domestic stockholding and a reduced reliance on world markets to make up domestic shortfalls. The coefficients of variation in Table 2.2 seem to indicate, for whatever reason, that there has been at least a relative shift in Soviet grain management practices.

The Effects on United States Wheat and Corn Markets

From a U.S. perspective, the stabilization of Soviet grain imports was only an intermediate objective of the first long-term agreement. It was hoped that more stable Soviet purchases would ultimately lead to more orderly U.S. grain markets (i.e., less price variation) than had prevailed in the early 1970s.

Examination of the variation in the price of U.S. futures contracts provides a rudimentary measurement of the impact of the LTA on U.S. grain markets. Ordinary least-squares was used to explain the quarterly varia-tion in wheat and corn futures contracts from the third quarter of 1972 through the second quarter of 1982. The quarterly range of futures prices for contracts coming due in the next quarter on the Chicago Board of Trade was used as the measure of price variability. Futures price ranges were used rather than changes in spot prices because the former would be more appropriate for capturing market expectations.

Price movements in futures contracts can be gener-ated by a multiplicity of factors. White noise, or unexplained movements, account for a major proportion of the variation since traders will frequently bid up or bid down daily prices based on rumors of potential changes in supply or demand. Hence, it is not suffi-cient to explain futures contracts price variability with changes in fundamental supply and demand components since expectations also play an extremely important role.

The equations in Table 2.3 attempt to measure the importance of the U.S.-Soviet grains agreement along with other factors which are likely to be significant in determining price variability of wheat and corn futures contracts. The influence of the grains agreement is represented by two sets of dummy variables each of which represents an alternative hypothesis about how the agreement affected the market. One of the dummies, BTAI, reflects a change in market structure or behavior beginning in the third quarter of 1976 when the trade agreement was implemented. If this variable is negative and a significant explanation of price variation, then there is reason to believe that the trade agreement reduced price variability by reducing the variability in U.S.-Soviet grain trade. The other dummy variable, BTAA, reflects a change in market structure or behavior beginning a year earlier, i.e. in the third quarter of 1975, when the agreement was first announced. If BTAA has a significant inverse relation with price variabil-ity, then it would appear that the announcement of the trade agreement resulted in a change in the behavior or expectations of market participants in a way that

TABLE 2.3
The Effect of the 1975 U.S.-Soviet Long-Term Agreement on Quarterly Price Variations
in Wheat and Corn Futures Contracts (Student's t statistics are in parentheses)

Dependent Variable: Quarterly Wheat Price Range of Wheat Futures Contracts

Equation	Constant	BTAA	BTAI	GX	USWST	R^2	DW
I	161			.002 (1.37)	-3.42 (4.29)	.38	1.76
II	128		-25 (1.15)	.003 (1.74)	-2.73 (2.75)	.40	1.77
III	124	-40 (2.26)		.003 (2.21)	-2.64 (3.18)	.46	1.85

Dependent Variable: Quarterly Corn Price Range of Corn Futures Contracts

Equation	Constant	BTAA	BTAI	GX	USCST	R^2	DW
I	71.82			.002 (1.86)	-.74 (2.45)	.15	1.58
II	28.85		-33.06 (2.55)	.003 (2.68)	-.45 (1.48)	.28	1.82
III	43.41	-36.60 (3.24)		.004 (3.27)	-.71 (2.64)	.35	2.25

Definition of Terms:
BTAA -- Dummy variable signifying the announcement of the 1975 U.S.-Soviet LTA. (BTAA = 0
        from Q3 1972 to Q2 1975; BTAA = 1 from Q3 1975 to Q2 1982).
BTAI -- Dummy variable signifying the implementation of the 1975 U.S.-Soviet LTA. (BTAI=0
        from Q3 1972 to Q2 1976; BTAI = 1 from Q3 1976 to Q2 1982).
GX   -- Quarterly total exports of wheat and corn from the United States, Argentina,
        Canada, Australia, France, and Thailand (TMT).
USWST-- Beginning stocks of U.S. wheat, deseasonalized using XII procedure (TMT).
USCST-- Beginning stocks of U.S. corn, deseasonalized using the XII procedure.

reduced price variability. Whether actual U.S.-Soviet grain trade is more stable or not as a result of the trade agreement may not be important as long as futures market participants perceive the market as being less uncertain as a result.

A number of other variables were used in an attempt to explain the price variability of wheat and corn futures contracts. Grain exports of major exporters (GX) and U.S. wheat and corn stocks (USWST and USCST, respectively) adjusted for seasonal variations were the most successful. Although there are strong theoretical reasons for using first differences of these variables, the absolute levels of exports and stocks proved to be a more powerful explanation of price variability.

Two sets of three equations are shown in Table 2.3, one set for the quarterly range in wheat contract prices and one for the quarterly range of corn contract prices. The first equation in each set uses only grain exports and stocks as explanatory variables, the second equation includes the dummy for the implementation of the agreement (third quarter 1976), and the third equation uses the dummy for the announcement of the agreement (third quarter 1975). All the explanatory variables have the expected signs, i.e. the price variability of futures contracts is positively correlated with exports and negatively correlated with the level of U.S. stocks and the dummy for the U.S.-Soviet grains agreement. None of the equations can explain more than half the total variability. The third equations for both wheat and corn (with the dummy for the announcement of the agreement) have the highest R-square statistic.

Plots of the predicted values of each equation are shown against the actual range of price changes within the quarter for wheat and corn in Figures 2.4 and 2.5 respectively. For wheat, the inclusion of the dummy variable for the announcement of the agreement results (Equation III) tracks only somewhat better than Equation II which contains the dummy variable denoting implementation of the agreement. This is to be expected given the relative levels of the R^2 and "Student's" t statistics. The same comparison for corn, however, indicates that the announcement of the agreement had a greater relative impact on the variability of quarterly corn futures prices than did the implementation.

It is significant that grain exports and the level of U.S. stocks, which had begun to accumulate again in 1975 and 1976, do not fully explain the decrease in price variability which occurred in the 1975-76 period. The dummy for the trade agreement announcement adds a significant measure of explanation.

These findings tend to confirm the results of Table 2.2 which showed that the variability of Soviet grain imports declined only in a relative sense following implementation of the first U.S.-Soviet agreement.

54

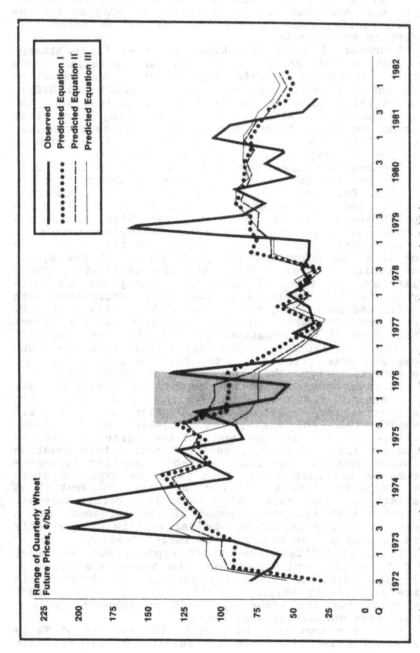

Figure 2.4 Estimates of quarterly wheat price variability.

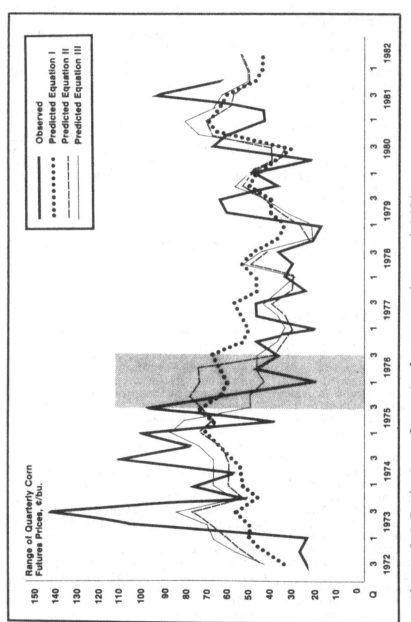

Figure 2.5 Estimates of quarterly corn price variability.

Although this may be important over time, it is not
likely to have any immediate effect on futures prices
except and unless these relative changes are implicitly
reflected in traders' expectations about how the market
will function as a result.

The results appear to be a statement about factors
which affect the nervousness of U.S. grain futures markets. Certainly the fact that the announcement of the
agreement proved to be a better explanation of the
decrease in price variability than did the actual implementation, indicates that traders' perceptions of the
riskiness of the market environment can often be just as
important as changes in supply and demand factors.
Traders may have correctly perceived the trade agreement
not only as a limit on Soviet actions, but also as a
constraint on the ability of the United States to impose
an embargo on grain shipments to the Soviet Union.
Given the grain trade environment of the early 1970s,
the 1975 U.S.-Soviet agreement was probably significant
in institutionalizing Soviet grain exports which had
earlier been a major source of market disruptions. As
such, it reduced both the perceived and actual price
risk to grain traders.

CONCLUSIONS

The 1975 U.S.-Soviet grains agreement expired in
1983 after two one-year extensions. It was replaced by
a new five-year agreement beginning October 1, 1983
which specified a minimum annual Soviet purchase of nine
million tons of grain. Of this amount, at least four
million tons each of wheat and corn must be purchased.
The agreement allows the Soviets to increase purchases
to twelve million tons without consultations and permits
the substitution of one ton of soybeans and/or soybean
meal for two tons of grain providing the minimum
purchase requirements for wheat and corn are met.

The evidence showing the stabilizing influence of
the original U.S.-Soviet agreement is circumstantial and
based heavily on the role the agreement played within
the context of world grain trade in the 1970s and very
early 1980s. The first agreement was signed at a time
of short world grain supplies, high and volatile grain
prices, and uncertainty about the longer term role of
the Soviet Union in world grain markets. The second
agreement was signed in August 1983 following a period
of accumulating U.S. grain stocks, falling prices, and
greater certainty that the Soviet Union would be a long
term major importer of grains.

The change in market environment, however, has
created new pressures for the agreement. The purchase
requirements carry implicit minimum prices equal to the
U.S. loan rates on wheat and corn. This became

important in the second year (1984-85) of the current
agreement when the Soviet Union chose to purchase only
2.9 of its 4 million ton annual wheat obligation. A
good Soviet harvest and large price discounts on Argentine wheat (which the United States could not match)
made the purchase of the entire 4 million tons of wheat
more costly than the Soviets were willing to accept.
The Soviet Union, however, did purchase 15.8 million
tons of corn from the United States in the 1984-85
agreement year. This experience demonstrates the
fraility of long term agreements in the face of strong
market forces.

The question arises whether the second agreement
can, in a much different trading environment, play a
role similar to the first agreement, and if not, what
role can the second agreement play which would be of the
greatest long term benefit to the United States?
Despite the failure of the Soviet Union to fulfill its
wheat purchase requirements, the spirit of the agreement
continues to be in the long term interests of both
nations. However, without a change in the world market
environment or a decline in U.S. loan rates, adjustments
in the obligations of the agreement may eventually need
to be made if the spirit of cooperation in grain trade
is to be maintained.

The change in world market conditions and the
experience with the first agreement has altered the
objectives of the second trade pact. The grain sales
suspension by the United States has heightened awareness
of the potential for events unrelated to the grain trade
to disrupt U.S.-Soviet trade flows while concern about
grain availability has abated. The stipulation in the
first agreement that the Soviet Union could increase its
annual purchases from six to eight million tons without
consultation provided U.S. grain supplies were in excess
of 225 million tons has been eliminated from the second
accord. They are merely to: 1) "employ its good offices
to facilitate and encourage...sales by private commercial sources...", and 2) "...not exercise any discre-
tionary authority available to it under U.S. law to control exports purchased for supply to the Soviet Union"
...under the agreement. Except for this change, U.S.
government obligations are largely the same in the 1983
agreement as they were in the 1975 agreement.

Earlier analysis showed how bilateral agreements can
be an impediment to adjustments in world and domestic
markets by preventing trade from falling below the level
of a country's bilateral commitments when a change in
market prices would warrant such an adjustment. World
price variability is increased when bilateral agreements
impede these adjustments. These occurrences are rare.
More frequently, bilateral agreements merely redirect
trade flows without having any major impact, positive or
negative, on prices and price variability. The 1975

58

U.S.-Soviet agreement is, perhaps, the single case where a bilateral agreement has had a positive influence on the reduction of world grain price variability. It is unique to the circumstances surrounding U.S.-Soviet grain trade. The negotiation of a second five-year agreement reflects a recognition within both countries of these unique circumstances along with a realization that continued unimpeded grain trade between the United States and the Soviet Union is in the best interests of both nations despite whatever other differences they may have.

NOTES

1. Unwritten agreements will not be considered in this paper because (a) it is difficult to verify their existence and (b) the informal nature of these agreements probably limits their influence on world trade.
2. The analysis for an exporting country is parallel and will not be discussed here.
3. The model implicity assumes that the costs of transshipping grain from the Soviet Union to markets outside Eastern Europe would be considerably higher than for Soviet-Eastern European trade. The Soviet Union would therefore pursue internal domestic adjustments, in response to a large exportable surplus, rather than exporting to non-Eastern European markets.
4. This approach was originally used by Clifton Luttrell (1981).
5. The USDA estimates of Soviet stock levels are, at best, informed conjecture about what Soviet stocks should be given information and estimates of other balance sheet components. Hence the statistics for beginning stocks in Table 2.2 may reflect major errors in data.

REFERENCES

Chicago Board of Trade. Statistical Annual of Cash and Futures Data 1982.
Cochrane, Willard W. and Mary E. Ryan, American Farm Policy 1948-1973, University of Minnesota, Minneapolis (1976).
Economic Research Service. U. S. D. A. Feed Outlook and Situation Report. Various issues.
_____. U. S. D. A. Wheat Outlook and Situation Report. Various issues.
_____. U.S.D.A. "Major Developments in the U. S. Export Reporting System," Wheat Situation, WS-230 (November 1974): 10-11.

Hilker, James H. and Bob F. Jones, A Stochastic Simulation Analysis of the U.S.-USSR Grain Purchase Agreement, Agricultural Experiment Station Bulletin No. 356, Purdue University (December 1981).

Jones, Bob F. and Alan J. Webb. "Economic Implications of Renewal or Non-Renewal of the U.S.-Soviet Union Bilateral Trade Agreement" presented at the AAEA Symposium, "Issues in U.S./Soviet-Bloc Agricultural Trade in the Eighties," Clemson University (July 27, 1981).

Longmire, James and John Dunmore, Sources of Recent Changes in U.S. Agricultural Exports, ERS Staff Report No. AGES831219 (January, 1984).

Luttrell, Clifton B. "Grain Export Agreements -- No Gains, No Losses" Review. Federal Reserve Bank of St. Louis, Vol. 63 No. 7 (Aug./Sept. 1981): 23.

Seeley, Ralph, Price Elasticities from the IIASA World Agricultural Model, ERS Staff Report No. AGES850418 (May, 1985).

Tyres, Rodney, Agricultural Protection and Market Insulation: Analysis of International Impacts by Stochastic Simulation, Pacific Economic Papers No. 111, Australia-Japan Research Center (May, 1984).

White House Press Secretary, "White House Fact Sheet, Grain and Oil Trade Agreements with the U.S.S.R.," Office of the White House Press Secretary (October 20, 1975).

3

Harvest Failures in Eastern Europe: Planners' Responses and Their Implications for World Grain Markets

Josef C. Brada

INTRODUCTION

During the 1970s, the emergence of the Soviet Union, Poland and several of the other centrally planned economies of Eastern Europe as important purchasers of grain on the world market raised a number of questions for traders, government policy-makers and for economists. These questions center on two related issues. The first of these concerns our ability to predict, at least in some rough way, the volume and timing of grain purchases by socialist countries. In order to do this we need a model or conceptual framework on which to base the empirical research that is required for forecasting. Although a great deal of empirical research has been undertaken to estimate import demand functions for market economies, to carry over such a methodology to the analysis of grain imports by planned economies is to miss those features of the planned economy that make its import behavior both unique and, simultaneously, difficult to predict. The second issue surrounding the grain imports of the socialist countries is that of national welfare. The variability and unpredictability of socialist grain imports is viewed as having potentially negative consequences for grain exporters. First, sharply fluctuating imports by the socialist countries can be seen as transferring the costs of output instability from the socialist importer to the exporting country. Thus Soviet consumers of grain do not suffer from variations in the Soviet harvest because imports mitigate the consequences of such fluctuations and simultaneously pass them on to, say, American consumers and growers of grain in the form of sharply fluctuating grain prices and exports. Secondly, the variable nature of socialist imports makes the prediction of socialist demand for grain difficult, and thus may be a potential advantage for socialist importers when negotiating contracts with Western exporters.

In this paper both these issues are addressed. The first part of the study provides an analysis of the pattern of grain imports of five centrally planned economies. The objective is not to undertake an econometric analysis of such imports, but rather to test a theory of planners' behavior against some observed empirical regularities. The second part of the paper then turns to policy issues and explores how grain exporting market economies can protect themselves against the costs of demand irregularity without foregoing the benefits of trade with the planned economies.

THE DEMAND FOR GRAIN IMPORTS

Planners' Behavior and the Demand for Grain

In a market economy the decisions that are ultimately reflected in the volume of grain imports are made by many independent agents who seek to maximize their welfare in the face of limited resources, incomes and information regarding the future. In a planned economy, while similar circumstances can lead to many of the same types of behavior, it is important to recognize that such behavior is not initiated by agents in response to price signals, but rather that such behavior reflects policy decisions made by planners. Thus any effort to forecast the grain imports of the planned economies must rest not on the "law of large numbers" as it does in the case of a market economy but rather on our ability to explain and predict the behavior of a small number of planners.

Planners may be viewed as performing two functions; the allocation of resources so as to promote economic growth and the allocation of resources so as to maintain some measure of equilibrium between the supply of, and requirements for individual commodities. The former activity relates to the long-term, the latter to the short-term. For example, the growing volume of East European grain imports during the 1970s reflects a long-term decision to increase the supply of meat without simultaneously increasing the output of animal feeds to the same degree. The implication of such a long-term policy decision for grain imports is not difficult to predict, nor to model econometrically. Unfortunately all that such econometric estimates can provide is an indication of the trend of socialist grain imports. Of greater importance to the trader and policymaker are the deviations from the trend. In particular we wish to understand how grain imports will change relative to the trend when there are poor harvests in the planned economies. An understanding of such deviations from the trend requires an understanding of how planners choose

to satisfy their second role -- that of equilibrators
of supplies and requirements in the short-run. Consequently we must separate long-term, or trend-deter-
mining, factors from short-term, equilibrating policies
in order to gain some understanding of how planners
choose the volume of grain imports.

Planners, of course, must strive for equilibrium,
but the ways of achieving it in the long-run are relatively limited. Thus a decision regarding the growth
rate of meat production and the volume of resources to
be devoted to crop production yields a relatively
inflexible option regarding the trend of feed imports.
By way of contrast, in the short-run there is much
greater flexibility because the planners are compelled
either to discard long-term objectives temporarily or to
violate long-term physical or biological relationships.
For example in the case of a poor harvest, planners can
reduce either the size of herds or import more grain, in
either case deviating temporarily from long-term
trends. Alternatively they may attempt to use up stocks
of grain, or to feed less to an unchanged number of
animals thus stretching available supplies. These
alternatives represent a way of violating, in a temporary way, the long-term physical and biological relationships between crop and animal output. The problem for
the forecaster is then to determine precisely how planners will choose to violate these long-term relationships.

How planners react to poor harvests depends on a
number of factors; the extent and nature of the harvest
failure, the current state of the economy, the weight
they assign to various objectives, and their expectations regarding the future. In the next section of this
chapter we examine how planners in Bulgaria, Czechoslovakia, Hungary, Poland and Romania have reacted to a
total of nineteen poor harvests in the period from 1960
to 1980.

East European Agricultural Crises, 1960-1980

When crop production falls, planners must make
adjustments in supply and requirements. As a general
tendency, the amount of grain given over to human consumption has not been permitted to fluctuate greatly
around its downward trend. Thus the principal decision
for the planners must involve the relationship between
grain imports and changes in animal herds. These
changes, however, influence meat production in both the
short-run and the long-run.

Because the supply of meat to the population has
become a particularly important indicator of economic
welfare both to planners and to consumers in the 1970s,
we should expect that in the 1970s planners were more

willing to import grain to maintain animal herds and
meat to make up for shortfalls in domestic production
than they were in the 1960s. Against this we must weigh
the increasing balance-of-payments concerns faced by
planners towards the end of the decade and their resultant need to conserve convertible currency.

Types of Harvest Failures. The nineteen episodes of
harvest failure considered in this paper were selected
on the basis of yield data for wheat and for one other
major grain crop; corn in Bulgaria, Hungary and
Romania; barley in Czechoslovakia; and rye in
Poland.[1] The sample period was divided into four
five-year periods, 1961-65, 1966-70, 1971-75, 1976-80.
A year would be included among the episodes examined if
in that year the yields for the two crops were both less
than the yields for the previous year and also less than
the average yield for the five-year period. In addition, certain episodes were included where there was,
for only one crop, a very large decline in yield both
from the previous year and relative to the five-year
average yield. Because each episode of poor harvests
involves adjustments in meat production, grain imports
and animal herds, our analysis of each episode covers
three years -- the year of the poor harvest and two
subsequent years when the effects of the adjustment work
themselves out. For those episodes where adjustment was
complicated by further harvest failures in the subsequent years, the period of analysis was extended. A
summary of the nineteen episodes is contained, by country, in Tables 3.1-3.5. For each episode or case, we
report changes in crop yields, in imports of grains and
meats, in meat production, and in the number of cattle,
hogs and poultry. To facilitate the analysis of these
cases they were divided into three types.

Type 1: One poor harvest followed by recovery.
There were five cases of this type, one in Bulgaria, two
in Czechoslovakia, one in Hungary and one in Romania.
In three cases, there was a decline in yields of both
crops; Czechoslovakia, Cases 2 and 3, and Romania, Case
2. In two cases, Bulgaria, Case 2, and Hungary, Case 3,
there was a decline in only one crop.

Since poor harvests are by their nature exceptional
occurrences, one might assume that planners facing a
poor harvest form an expectation that such abnormal circumstances will not arise again in the following one or
two years and that what they must do is to return the
agricultural sector back to equilibrium on the assumption that the subsequent two years will be characterized
by at least normal crop yields.

Given these expectations on the part of the planners
we expect them to deal with the shortfall in domestic
production in the following way. Imports of grains will
be increased either in the year of the poor harvest, or
in the subsequent year. The delay in increasing imports

TABLE 3.1
Consequences of Poor Harvests in Bulgaria, 1960-80.

Percentage change in:

Year	Wheat yields vs.		Corn yields vs.		Imports of		Meat Production	Numbers of		
	Previous Year	Five-Year Average	Previous Year	Five-Year Average	Grains	Meat		Cattle	Hogs	Poultry
					------- C A S E 1 -------					
1961	-18	-15	-5	-11	-10	-67	15	3	-9	-2
1962	8	-9	7	-5	51	-70	16	0	-11	-8
1963	-5	-13			87	438	-7	-6	2	5
1964	11	-3			-76	-17	9	-1	24	0
					------- C A S E 2 -------					
1965			-29	-12	-51	124	14	-2	-8	-5
1966			-10	-5	+30	-44	6	-4	-5	13
1967					-90	202	3	-2	2	17
					------- C A S E 3 -------					
1968	-22	-12	-9	-17	854	-44	9	-5	-10	-10
1969	+3	-10			31	-22	-6	-3	-8	19
1970					-70	44	-16	2	20	14
					------- C A S E 4 -------					
1974			-25	-21	484	258	2	7	42	-5
1975					2	-52	14	7	14	8
1976					-34	-15	13	4	-11	4
					------- C A S E 5 -------					
1977			-14	-11	-55	-30	-3	1	-2	4
1978					325	-28	2	1	11	-2
1979	-6	-7			31	-70	8	1	2	2

Source: Calculated from United States Department of Agriculture, Agricultural Statistics of Eastern Europe and the Soviet Union, 1960-80 (Washington: USDA, 1983) and FAO, Yearbook (Rome: FAO, various years).

TABLE 3.2
Consequences of Poor Harvests in Czechoslovakia, 1960-80.

Year	Wheat yields vs. Previous Year	Wheat yields vs. Five-Year Average	Barley yields vs. Previous Year	Barley yields vs. Five-Year Average	Imports of Grains	Imports of Meat	Meat Production	Number of Cattle	Number of Hogs	Poultry
				---- C A S E 1 ----						
1964	-10	-10	-11	-8	33	-30	7	-1	5	-4
1965			0	-8	-30	-24	9	-1	-10	-4
1966					-12	60	-3	+2	-4	6
1967					-10	-1	6	-1	6	6
				---- C A S E 2 ----						
1972	-5	-7	-7	-8	-23	-42	6	3	3	2
1973					2	153	1	2	3	5
1974					-26	-60	1	0	7	-4
				---- C A S E 3 ----						
1975	-10	-2	-18	-6	-17	-51	3	2	-1	2
1976					113	-38	-2	2	2	10
1977					-43	75	4	3	10	1

Source: See Table .1.

TABLE 3.3
Consequences of Poor Harvests in Hungary, 1960-80.

| | Wheat yields vs. | | Corn yields vs. | | Percentage change in: | | | | | |
	Previous Year	Five-Year Average	Previous Year	Five-Year Average	Imports of Grains	Meat	Meat Production	Cattle	Numbers of Hogs	Poultry
Year										
					---- CASE 1 ----					
1961	14	3	-19	-23	80	-20	8	2	9	NA
1962	-6	-4	24	-4	9	2	4	-3	-10	NA
1963	-12	-16			-2	83	-1	-5	-2	NA
1964					-12	18	0	+5	-2	NA
1965					-12	-22	5	-1	-6	NA
					---- CASE 2 ----					
1967			-10	-12	-21	87	2	4	9	2
1968			5	-8	67	-68	11	-2	-13	-5
1969					-25	-14	-4	-5	-2	26
					---- CASE 3 ----					
1970	-21	-12			-40	165	0	-1	28	8
1971					269	-78	11	-2	4	-4
1972					-4	-28	4	+1	-10	-14
					---- CASE 4 ----					
1975	-15	-4	-24	-22	-55	-40	6	-5	-16	-3
1976			+22	-4	36	133	-1	-1	13	13
1977					21	-66	6	3	0	2
1978					37	-40	0	1	2	3

Source: See Table .1.

68

TABLE 3.4
Consequences of Poor Harvests in Romania, 1960-80.

Percentage change in:

Year	Wheat yields vs. Previous Year	Wheat yields vs. Five-Year Average	Corn yields vs. Previous Year	Corn yields vs. Five-Year Average	Imports of Grains	Imports of Meat	Meat Production	Numbers of Cattle	Numbers of Hogs	Poultry
					---- CASE 1 ----					
1962	-1	-8	-5	-10	36	-32	10	-3	-3	-24
1963	-1	-9			152	-80	-21	1	3	12
1964	-2	-11			637	136	17	3	30	4
1965					-91	-100	6	4	-11	0
					---- CASE 2 ----					
1970	-8	-14	-9	-5	575	53	2	4	6	1
1971					148	492	3	6	22	13
1972					-15	-6	17	4	13	5
					---- CASE 3 ----					
1973	-10	-6	-19	-7	-48	-75	13	2	2	3
1974	-1	-7	1	-6	472	-17	7	1	-5	2
1975					-11	-71	2	2	3	16
1976					34	249	7	4	17	16
					---- CASE 4 ----					
1977			-11	-9	-24	52	6	-1	4	-3
1978			5	-5	-4	120	1	3	6	12
1979	-19	-17			78	49	10	0	5	4
1980					27	62	-4	0	-8	-5

Source: See Table .1.

TABLE 3.5
Consequences of Poor Harvests in Poland, 1960-80.

	Wheat yields vs.		Rye yields vs.		Percentage change in:			Numbers of		
	Previous Year	Five-Year Average	Previous Year	Five-Year Average	Imports of Grains	Imports of Meat	Meat Production	Cattle	Hogs	Poultry
Year										
					------ C A S E 1 ------					
1962			-17	-13	-12	-66	3	1	-2	-3
1963	-6	-5			32	1742	-5	5	-10	5
1964			-4	-4	-5	-24	1	1	15	3
1965					-3	-1	9	4	1	-1
1966					-33	42	4	-3	2	1
					------ C A S E 2 ------					
1970			-19	-13	31	26	-9	-10	-6	2
1971					18	250	1	3	22	1
1972	-5	-11			5	-64	12	6	12	5
1973					4	3	10	9	13	1
1974					25	-89	12	4	1	3
					------ C A S E 3 ------					
1977			-15	-7	-6	124	0	3	23	6
1978					28	-69	10	0	3	0
1979	-17	-8	-26	-16	-1	-95	4	-2	-1	1
1980	-4	-11	19	0	6	-323	-3	0	-8	-5

Source: See Table .1.

may occur either because the factors leading to the poor harvest may only become evident during the harvest season thus leaving little scope for increasing imports in the current calendar year, or, alternatively, imports may be delayed for balance-of-payments reasons. A strategy for conserving foreign exchange might dictate delaying imports in order to obtain a more favorable price (or, of course, speeding up imports to take advantage of depressed prices) or waiting to see whether next year's harvest might not be particularly bountiful, thus reducing the quantity that would have to be imported. Clearly the volume of grain stocks that can be drawn down and the possibilities for reducing animal herds interact with the availability of convertible currencies to influence planners' decisions.

The decision regarding grain imports in turn makes its influence felt on the size of animal herds and on meat production. To the extent that animal herds are reduced, current meat production may increase as animals are slaughtered. The number of cattle is least likely to fluctuate, because cattle herds can only be rebuilt slowly and also because cattle are less dependent on grain for feed than are hogs and poultry. The latter two types of animals are more likely to fluctuate in numbers both because they have the disadvantage of being more reliant on grain and the advantage of more rapid reproduction than cattle.[2]

Although a decision to reduce herd size may spur meat output in the short-run, the output of meat is likely to fall in the future as the stock of animals will be smaller and must be rebuilt. Moreover, while distress slaughtering may stimulate meat production in the short-run, insufficient rations for animals may lead to a decline in meat production even as the number of animals slaughtered is increasing. If planners are concerned about fluctuations in the supply of meat to the population then we would expect to see a negative relationship between meat production and meat imports.

In three of the cases planners' responses were qualitatively similar. These are Bulgaria, Case 4, from 1974 to 1976; Czechoslovakia, Case 3, from 1975 to 1977; Hungary, Case 3, from 1970 to 1972. In the cases of Czechoslovakia and Hungary, grain imports were reduced in the first year but then increased very sharply in the second year. In part this may reflect the fact that stocks of grain must have been relatively high since in both countries the decline in yields from the previous year is greater than the decline from the five-year average, indicating that the harvest prior to the start of the crisis must have been quite good. In Bulgaria this was less the case, and perhaps as a result of lower grain stocks, imports of grain increased sharply in the first year and then remained at that level for the second. In all three countries grain imports were reduced in the third year, by 34 percent in Bulgaria and Czechoslovakia but only by 4 percent in Hungary. Although all three countries followed a similar pattern in terms of grain imports, changes in animal herds tended to be more country specific. In both Bulgaria and Czechoslovakia the number of cattle continued to increase over the entire three-year period. In contrast, in Hungary the number of cattle declined slightly. In the case of hogs, there appeared to be a rather perverse reaction in Bulgaria, with a 42 percent increase in numbers and in Hungary with a 28 percent rise. In both countries this represented a major policy shift in that these increases represent a long-term, rather than temporary, increase in the number of hogs maintained. In Czechoslovakia more moderate rates of growth are evident. In the case of poultry, there are no similarities; in Bulgaria the numbers decreased in the first year and then increased; while in Czechoslovakia they increased each year. In Hungary, the pattern was the opposite of that in Bulgaria.

Different policies toward livestock inventories also led to differences in meat production. In Czechoslovakia, meat production increased by 3 percent in the first year, but then stagnated, reflecting both the slow rate of increase of livestock and possibly reduced rations. To offset the slowdown in meat production, imports of meat were increased in the second year, but scaled back sharply in the third year. In Bulgaria and Hungary the stagnation of meat production in the first year led to sharp increases in imports of meat followed by large decreases in meat import in subsequent years when meat production increased at very rapid rates. To the extent that domestic meat production can be replaced by meat imports, such imports can be viewed as a deferral of and substitute for current imports of grain. However, it should be noted that despite the large percentage increases in meat imports, such imports made up for only part of the shortfall in domestic production.

Two cases do not fit the above general pattern. The first of these is Czechoslovakia, Case 2, where a moderately poor harvest appears to have caused no great upsurge in grain imports. Rather the planners appear to have settled for very modest increases in animal herds over the three-year period. As a result of the slow growth of herds and possibly of reduced rations, meat production stagnated after the first year and, while imports of meat increased in the second year, they declined sharply in the third despite little progress in expanding the domestic production of meat. Overall, compared to the previous cases, this represents a rather autarchic approach to dealing with harvest shortfalls.

The second case represents somewhat the opposite reaction on the part of planners. In Romania, Case 2, a poor harvest apears to have triggered a set of responses aimed not at using the foreign trade sector to mitigate the effects of a poor harvest but rather at overcoming these effects altogether. Grain imports increased sharply in the first two years and did not fall much in the third. Animal herd growth rates were maintained at high levels and meat production was supplemented by rapid expansion of meat imports, albeit from a low base.

Type 2: Two consecutive poor harvests followed by recovery. There were eight cases of this type: Bulgaria Cases 1 and 3, Czechoslovakia Case 1, Hungary Cases 1, 2 and 4, Romania Cases 1 and 3. In a situation where a second poor harvest follows immediately upon the first, the planners face a much more difficult situation in the second year. By then all the easy solutions have been tried -- stocks of grain have been reduced, animals have been held on short rations and planners must now make more difficult choices between increased reliance on imports of grain and a reduction in animal herds.

To highlight the difference between planners' behavior in the 1960s and the 1970s we examine cases of Type 2 by decade. In the 1960s there were six cases. The first year responses are neither qualitatively nor quantitatively different from those observed above for Type 1. However after that the pattern shifts. In three cases, Czechoslovakia, Case 1, and Hungary, Cases 1 and 2, after an effort to offset poor domestic harvests through grain imports in the first or second year, in the face of a second poor harvest, grain imports decline by the second or third year and continue at low or falling levels. Apparently in these cases planners were either unable or unwilling to maintain requisite grain imports for a second year, and chose to reduce herds instead. In Hungary, Cases 1 and 2, for example, in the second and third years cattle and hogs both experienced large declines in numbers that lasted well into the recovery period. In Czechoslovakia the pattern was similar. The effect on meat production was also similar. Increases in the first and particularly the second year, the latter evidently distress slaughtering, were followed by declines in meat production. Nor were the fluctuations in meat production offset by meat imports to the degree evident in Type 1 harvest fail-ures. Indeed in a number of years there was a direct rather than inverse relationship between meat production and meat imports, indicating that planners were forced to abandon both long-term objectives regarding the growth of herds and those regarding supplies of meat to the population.

The other three cases in the 1960s, Bulgaria 1 and 3 and Romania 1 reflect a different pattern of grain imports. Here the imports of grain increased in response to the worsening domestic situation. Nevertheless, except in the case of Romania, animal herds were reduced quite sharply. Romanian poultry numbers declined, but cattle increased modestly and hogs quite sharply. In Romania meat production exhibited erratic growth rates, which were not offset effectively by changes in meat imports, and meat supplies thus stagnated over the crisis and recovery period. Swings in meat production were also quite large in Bulgaria, though evidently efforts to use imports of meat to offset production fluctuations were more effective.

Although qualitatively there appear to be two types of response to harvest failures of Type 2 during the 1960s, the common element in these is that under the pressure of two poor harvests planners generally were forced to cut herd sizes quite sharply and supplies of meat to the population had to be curtailed. Thus while grain imports in some cases were used to ameliorate the effects of harvest failures, in the Type 2 cases the shortfall in grain output was not entirely or even adequately supplemented by imports.

The two harvest failures of this type in the 1970s, Hungary, Case 4 and Romania, Case 3 reveal a somewhat different pattern of behavior. In the case of Hungary after a decline in grain imports in the first year, grain imports rose steadily over the remainder of the period despite the recovery in wheat harvests after the first year and the recovery in corn production by 1977. As a result of these grain imports, moderate growth of animal herds was maintained. While cattle numbers fell sharply in the first year, by the third year growth was resumed. Pork and poultry also declined in numbers during the first year but thereafter rebounded in the second year and continued to increase slowly. The producton of meat increased quite sharply in 1975, reflecting the sharp decline in animal herds in that year. The decline in production of meat in 1976 was partly offset by an increase in meat imports and as meat producton increased in subsequent years imports of meat were reduced.

In Romania, a somewhat similar pattern may be observed in the 1973-76 period. Grain imports fell in the first year, but then increased very sharply and remained around the level of the second year for the remainder of the period. As a result of these imports, despite a three-year period of mediocre crop production, animal herds increased moderately except for hogs, which nevertheless held their own. As a result of this steady growth in animal herds the production of meat exhibited an erratic but nevertheless upward tendency, permitting the planners to reduce imports of meat through the 1973-75 period. It is also worth noting that the return to normalcy in crop production in 1976 led to a large growth in the numbers of animals, reflecting both planners' efforts to promote the long-term growth of meat production and the constraints that the failure of the harvests in 1973-75 imposed in this growth.

The comparison of the pre- and post-1970 Type 2 harvest failures clearly shows the role of planners' preferences in determining the volume of grain imports. In the earlier period planners were more willing to sacrifice the growth of animal herds in order to restrain grain imports at relatively low levels. In the 1970s planners were less willing to reduce the size of animal herds in the face of poor harvests and at the same time more willing to expend hard currencies for the grain imports required to maintain some, albeit slower than desired, growth in animal herds.

Type 3: One poor harvest with a second poor harvest later in the recovery period. Type 3 cases represent a situation intermediate between the two already discussed. To the extent that planners are afforded at least one good harvest after the initial bad harvest, there is some opportunity to restore equlibrium to the agrarian sector. Nevertheless, it is unlikely that in one year they could both repair the damage done in the previous one and establish a strong position from which to deal with another poor harvest.

As with the preceding discussion, we shall examine pre- and post-1970 experiences separately. Of the former there are two, Bulgaria, Case 2 and Poland, Case 1. In both cases, planners did not make very energetic efforts to employ imports of grain to offset the effects of domestic crop failures and, in fact, in both countries imports of grain fell over the period under examination. In Bulgaria, this had very negative consequences for animal herds. The number of cattle fell steadily during 1965-67 and the number of hogs, although increasing in 1967, was reduced by 12 percent. Only poultry numbers increased. As a result of the sharp reduction of animal herds in 1965, meat production increased sharply, and continued to increase, albeit more slowly in subsequent years as more animals were slaughtered.

In Poland a similar pattern of grain imports was implemented in 1962-66. Although there was an increase in imports in 1963 it was modest at best and subsequent years reflect a slow but steady decline. The consequences of this pattern of grain imports were less serious for animal herds in Poland than they had been in Bulgaria. Despite the decline in hogs and poultry in 1962-63, animal herds grew at modest rates. As a consequence, meat production also grew throughout the period, albeit erratically, but the only decline in meat production, in 1963, was offset by a sharp increase in imports of meat.

In general the post-1970 cases reflect the same sort of change in planners' behavior noted in the discussion of Type 2 cases. In Bulgaria, Case 5, a poor harvest in 1977 was followed by an increase in grain imports in 1978 and a further increase in 1979 in response to a decline in the grain harvest. As a result, cattle herds grew modestly as did the number of poultry. After a decline in numbers in 1977, hogs rebounded sharply in 1978 and continued to grow in numbers in 1979. While meat production did fall in the first year, it grew thereafter. The decline in meat imports throughout the period suggests that planners wished to conserve foreign exchange to permit greater grain imports rather than resorting to imports of meat as a short-term remedy.

The two Polish experiences in the 1970s, Cases 2 and 3, reflect similar planners' preferences under changing conditions. In Case 2 the government was committed to raising consumption levels and through reliance on foreign credits the foreign exchange constraint was overcome. Thus for the entire 1970-74 period grain imports increased. Consequently after a decline in numbers in 1970, cattle and hogs increased in numbers at very rapid rates. As a result, by 1972 meat production was also increasing rapidly, and meat imports were being reduced. When the planners faced a similar situation in 1977-80, the balance-of-payments constraint was much more of a concern. Thus although grain imports increased in 1978 there is no additional upsurge in imports in response to the poor harvests in 1979 and 1980. As a result, the growth in animal herds, aside from hogs in 1977, was quite slow and by 1979 herds were reduced, a process that accelerated in the years following our analysis. The production of meat reflects developments in the livestock sector, with slowing growth turning negative in 1980. Throughout, meat imports were being reduced after the increase of 1977, again reflecting concerns about the balance-of-payments and the need to maintain grain imports for the long-term.

The case of Romania in 1977-80 is somewhat unique in that despite a poor corn harvest in 1977 followed by a mediocre one in 1978 grain imports exhibited a negative trend. Thus it was not until 1978, in the face of a poor wheat harvest, that planners stepped up the volume of grain imports. As a consequence of this strategy, animal herds failed to show much dynamism, and meat production was erratic. Nevertheless, even in the face of increasingly severe balance of payments constraints the planners strove to maintain animal herds and their long-term objective of increasing the production of meat.

Responses to Agricultural Crises in the 1980s

The general pattern of planners' responses to poor harvests evident in the 1970s continued to dictate responses to similar harvest shortfalls in the 1980s

except in the case of Poland. In that country, balance-
of-payments considerations and the overriding need to
deal with the external debt as well as the breakdown of
the domestic economy forced planners to adopt entirely
different strategies.

While the shorter period available for analysis in
the 1980s precludes the use of the analytical structure
employed in the previous section, it is possible to
identify three poor harvests for analysis. Two of
these, Czechoslovakia and the German Democratic Republic, both in 1981 and the following years, displayed
quite similar behavior consistent with the pattern
observed during the 1970s. In Czechoslovakia wheat
yields declined by 10 percent in 1981 and remained
depressed in 1982. Barley yields were also low in 1981
and the total grain harvest was 7 percent lower in 1981
than in the previous year. Planners responded by reducing grain imports by 50 percent in 1981 but increased
them by 43 percent in 1982. Changes in animal herds
reflected the reduced availablility of feeds. The
largest response to feed shortages was in the number of
hogs, which fell by 8 percent in 1981, 3 percent in
1982, and 1 percent the following year. In contrast,
cattle continued to grow in numbers, increasing by 1-2
percent in 1981-1983. In part the rather steep decline
in the number of hogs represented a policy shift of a
longer term nature that coincided with the poor harvest
of 1981. During the 1960s and early 1970s, pork and
poultry were viewed as being the preferred means of
dealing with feed uncertainty in that their numbers
could be reduced and expanded in response to harvest
fluctuations. However, as wide fluctuations in animal
herds became less acceptable to policy makers, the
advantages of cattle came to the fore. Although the
numbers of cattle cannot be modified as quickly as those
of hogs and poultry, cattle are less dependent on grain
feeds, and thus can be carried through poor grain harvests by expanding the use of hay and naturally occurring forage, possibilities not as readily available in
the case of hogs and poultry. The decline in hog numbers increased meat production by 2 percent in 1981,
while reduced animal herds caused a 7 percent decline in
meat output in 1982. Meat imports reflected these
swings in production, falling by 33 percent in 1981 and
rising by 23 percent in 1982.

Events in the German Democratic Republic were caused
by very similar developments. Wheat yields were low in
1981 and 1982; barley yields were also low in 1981 and
the grain harvest in 1981 declined by 8 percent. Grain
imports declined in 1981 by 25 percent and by 32 percent
in 1982. Given the overall reduction in the
availability of grain, there was a general reduction in
the size of animal herds. Cattle numbers stagnated in
1981 and then declined by 1 percent, while hog numbers

decreased by 6 percent in 1982 as did poultry. The rather more drastic changes in herd size implemented by the GDR also caused wider fluctuations in meat production, which increased by 5 percent in 1981 and then fell by 8 percent in 1982. Meat imports corresponded to changes in production, falling by 25 percent in 1981 but increasing by 173 percent in 1982. Thus in response to relatively mild, Type 1, harvest problems, planners' responses in Czechoslovakia and the GDR were quite similar to responses observed in the 1970s reflecting a relatively stable set of goals for agriculture.

In Poland the effects of a poor harvest in 1980 were completely overshadowed by efforts to deal with that country's balance-of-payments difficulties. Thus grain imports were reduced by 8 percent in 1981 and by 40 percent in 1982. Since grain imports in 1980 had been equivalent to over 40 percent of domestic production this readjustment of imports required significant structural change in the agricultural sector. Animal herds clearly had to be reduced to reflect the decline in feed availability. Thus between 1980 and 1983 cattle declined by 12 percent, hogs by 16 percent and poultry by 20 percent. Meat production fell by 29 percent over the same period. Polish meat import policies were also the victims of the payments crisis, and, although they rose by 261 percent in 1981, they fell by 50 percent in 1982. More telling and significant in volume were the changes in Polish meat exports which declined by 50 percent in 1981 and 19 percent in 1982. The Polish case thus tells us little about responses to a poor harvest, largely because the change in availability of imported grains represented a structural change that forced the abandonment of past policies.

Implications for Prediction and Modelling

Although planners have a number of ways of dealing with agricultural crises, we have seen that in the 1960s and again in the 1970s and 1980s planners in all countries adopted qualitatively similar methods of coping. Although we have not examined all the causal factors in detail, it is evident that planners' preferences play an important role in determining how to deal with poor harvests. In the 1960s they were willing to let the animal sector and consequently the long-term growth of meat production bear much of the costs of adjustment. In the 1970s and 1980s there was a greater tendency to shield the livestock sector from the effects of crop failures. Thus prediction and modelling of East European grain import policies must take changes in planners' values and preferences into account. Moreover, decisions regarding appropriate responses to harvest failures reflect crisis decision-making and thus econometric models based on time series evidence may lead to misleading results since they mix periods of normalcy with periods of crisis. On the positive side, the similarity of behavior during crises in the countries examined suggests that cross-section studies of crisis behavior may yield useful econometric and predictive results.

IMPLICATIONS FOR WORLD GRAIN MARKETS

The evident variability of grain imports by the member countries of the Council for Mutual Economic Assistance (CMEA) and the rather inelastic nature of that import demand since the 1970s demonstrated in the foregoing section have induced a number of grain-exporting countries to enter into long-term agreements with these countries. In this section we examine the rationale behind such agreements and their implications for the world grain market.[3]

Trading with a Centrally Planned Economy

Market Power. The most pessimistic view of trade between the United States and a CPE views the latter as a shrewdly calculating, well informed trader exercising the market power of its state trading apparatus to extract all the possible advantages from trade. Raymond Vernon (1979) has presented a heuristic argument for such a scenario. More relevant from our standpoint is Batra's (1976) theoretical elaboration of Vernon's themes. In Batra's model, CPE planners conduct domestic production and international trade such that they maximize a planners' preference function: $U = U(X,G)$, where X is the composite export good and G grain, subject to domestic production possibilities and balanced trade. If both the CPE and the United States exercise some market power, i.e., if their terms of trade depend on their volume of trade, then we have the situation shown in Figure 3.1. OC_S represents the CPE offer curve, depicting desired combinations of exports of X and imports of G at various price ratios, P_X/P_g. OC_u represents desired combinations of exports of G and imports of X for the United States at various levels of P_X/P_g. If neither country tries to exploit its market power, trade will take place at point F with the United States exporting OG_F of grain and importing OX_F of the CPE's export good. Prices of the two goods in trade between the United States and the CPE are given by the slope of the line TT_F. The CPE's welfare is represented by the trade indifference curve U_1. Note that the shape of the indifference curve implies that the CPE's planners are willing to substitute between G and X in domestic consumption.

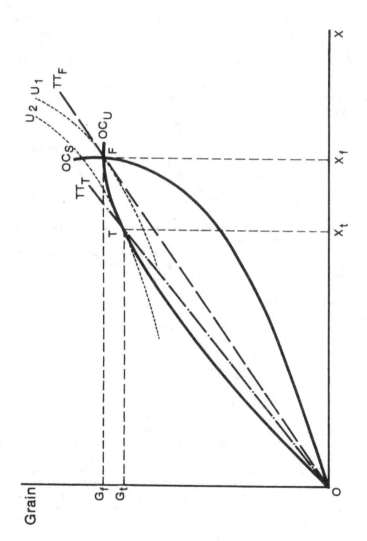

Figure 3.1 Traditional Offer Curve Analysis of Trade Between
a Market Economy and a CPE

The CPE traders, however, could increase their gains from trade at the expense of the United States by offering to exchange OX_T of X for OG_T of grain and thus trading at point T. By doing so they move to trade indifference curve U_2, the highest attainable indifference curve given the U. S. offer curve OC_U. While this is the familiar argument in favor of the optimal tariff, the CPE's economy is, in Batra's view, more likely to be able to reach point T than is a market economy such as the United States. This is because the state monopoly over trade enables the planners to choose to trade at T; no public announcement or implementation of internationally visible tariffs or quotas is necessary. Consequently the United States may not be aware of the fact that the CPE's planners are exercising international market power. Even if U.S. policymakers are aware that the CPE is skewing the gains from trade in their favor, retaliation is risky. By imposing a counter-tariff of its own, the United States might hope to shift the terms of trade in its favor. In Figure 3.2 this is shown by shifting the U.S. offer curve from OC_U to OC_U'. However, the CPE may choose to trade at point T', which maximizes its welfare, but worsens rather than improves the U.S. terms of trade. Thus, because the CPE's response to U.S. commercial policy changes is indeterminate, a priori, the use of market power by the United States is fraught with risk. Moreover, any attempt to exercise market power by the United States would require the imposition of tariffs or quotas which would be visible to the CPE and might also be in conflict with obligations of the United States under international agreements such as GATT.

Thomas Wolf (1978, 1982a) has criticized the Batra model as requiring assumptions that are not consistent with the reality of CPE planning and foreign trade decision-making. Specifically Wolf raises two objections. The first is that since the CPE's production is planned there is unlikely to be any significant response either in annual plans or within five-year plans to changes in the terms of trade. Thus the kind of production responses necessary for the implementation of Batra's analogue to the optimal tariff are not likely to occur in the real-world CPE economy. Secondly, Wolf disputes Batra's formulation of the planners' welfare function. In Wolf's view, planners are unlikely to be willing to substitute more of one good for less of another once production and consumption targets are set. While this may be an excessively inflexible view of planners' preferences, within the context of this paper it is rather apt. The import good in our case is grain; the composite Soviet export good consists in large part of minerals, fuels and other raw materials. It is unlikely that Soviet and East European consumers

81

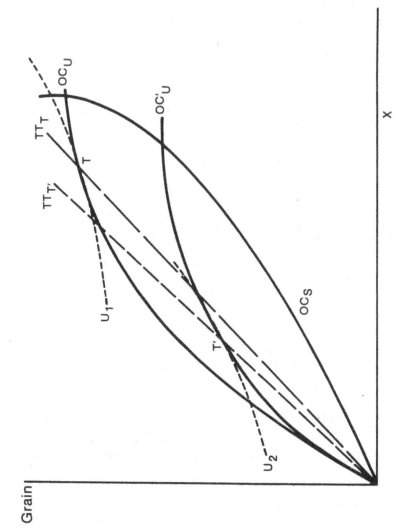

Figure 3.2   Offer Curve Analysis of the Exercise of Market Power
in Trade Between a Planned and a Market Economy

of livestock have much scope for substituting the latter goods for grain.

Wolf's model and its implications can be illustrated by means of Figures 3.3 and 3.4. In Figure 3.3 q'q represents the CPE's production possibilities. The line OC represents pattern of consumption desired by planners and reflects planners' preferences as illustrated by the indifference curves U_1, U_2. These indifference curves are L-shaped since by assumption planners are unwilling to substitute X for grain in consumption. The production point, P, does not change in response to changes in the terms of trade. Thus as the terms of trade improve from TT_1 to TT_2, consumption of both goods increases from C_1 to C_2 and welfare increases from U_1 to U_2. The offer curve corresponding to these assumptions is given in Figure 3.4 by OC_S.[4] With the U.S. offer curve represented by OC_U, trade takes place at point A , at prices given by TT_A. In view of the L-shaped CPE indifference curves, there is no benefit to planners in moving trade away from A. On the other hand, if the United States imposes a tariff, shifting its offer curve to OC_U', both its terms of trade, TT_B, and welfare must increase as trade takes place at B. Moreover, there is no incentive or possibility for the CPE to retaliate by exploiting its market power. Indeed, only if the U.S. offer curve is unit or inelastic, as shown in Figure 3.5, would it benefit the CPE to exploit its market power. It is worth noting that a U.S. policy subsidizing grain exports so as to keep the export price fixed would lead to a unit elastic offer curve and thus opportunity for the exploitation of CPE market power.

Bulk Buying. The CPE's state trading monopoly may act as a price taker on international markets within the context of the foregoing discussion, yet it may nevertheless seek to use its size to obtain better commercial terms from its western suppliers by means of bulk buying or long-term commitments. In making such efforts, the CPE grain monopoly probably does not differ from the state purchasing boards of other grain-importing countries. Consequently, such behavior is not discussed in this paper. A more serious possibility is that, because of secrecy surrounding its buying intentions the CPE can split up its purchases among several sellers in such a way as to keep the price of grain from rising until its purchases have been consumated.

This possibility is similar to Goldman's (1975) interpretation of the 1972 grain sales. American firms believe that the CPE's offer curve is given by OC_S in Figure 3.6. Terms of trade are established at TT_1. However, the true CPE offer curve is given by OC_S', indicating a greater CPE need for grain. If the true CPE offer curve were known to western suppliers, the

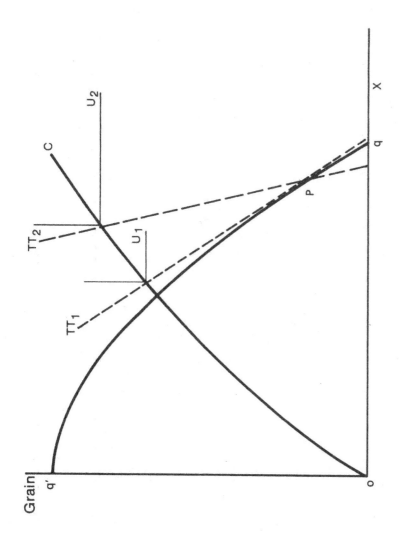

Figure 3.3 Production and Trade in a CPE with Fixed Consumption Proportions

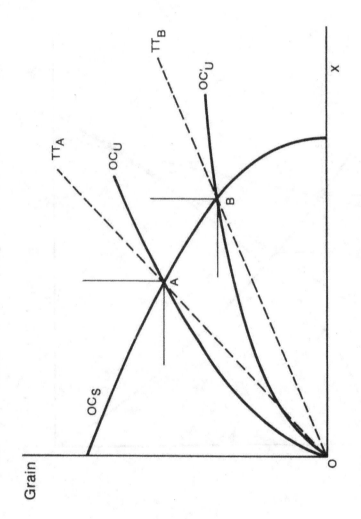

Figure 3.4 Wolf's Model of Trade Between the United States
and a Planned Economy

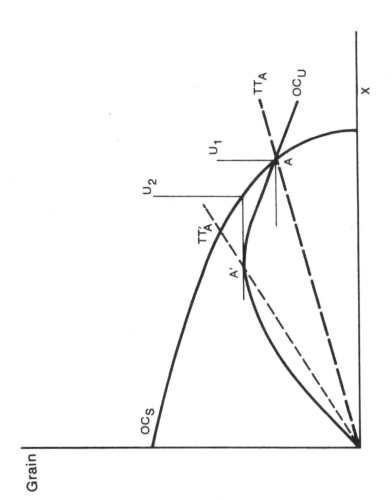

Figure 3.5 Gains to a Planned Economy Exercising Market Power

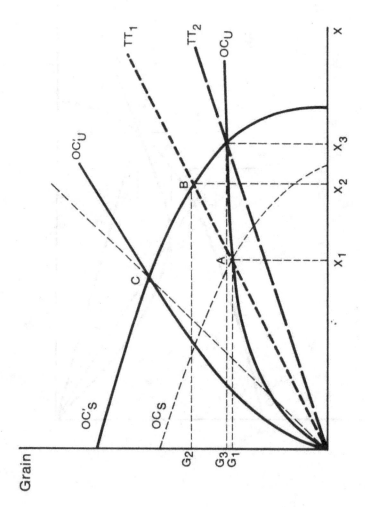

Figure 3.6 Benefits of Bulk Buying by the Planned Economy

price of grain would rise and the CPE would exchange OX_3 of X for OG_2 of grain. However, by approaching each firm in secret, the CPE grain monopoly signs a sufficiently large contract with each to purchase grain equal to OG_3. Since the participating firms and "the market" are thus unaware of the true volume of CPE purchases, they make their sales at prices TT_1 and the CPE exchanges OX_2 of X for OG_2 of grain. Only when the true volume of CPE demand is revealed do grain prices increase.

Such a technique for exploiting the advantages of a state trading monopoly are, however, somewhat limited. In order to be successful, the state trading monopoly must contract for all of its needs simultaneously and relatively early in the marketing year. While such early commitments may benefit the CPE in years in which world grain prices rise over time, it is also possible that good harvests in the West may cause the U.S. offer curve to shift upward, to OC_U'. In this case the CPE is locked into high prices at TT_1, when, by delaying they might have purchased grain at the more attractive prices TT_3. Johnson (1977, pp. 22-23) argues that this is precisely what occurred in 1975-76. The Soviets contracted for their grain needs earlier than other buyers and as a result paid higher average prices for grain than did importers who made their purchases over the course of the entire year.

We may thus conclude that the CPEs' tendency to contract early for their full import needs is not likely to be an important factor in shifting the benefits from trade in their favor. CPE traders tend to be risk averse and no doubt prefer securing needed supplies at known prices to the more speculative policy of making such purchases over time.

The Economics of Grain Agreements

In the foregoing section of this paper a number of issues have been raised in connection with the trade in grain between the United States and CPEs. In this section we evaluate how grain agreements negotiated by the United States with a number of CPEs deal with these potential problems. For the sake of ease of exposition we examine the 1976 U.S.-Soviet agreement although the analysis applies to the other agreements as well.

Turning first to the distribution of the gains from trade, the significant provision of the agreement is the one requiring the Soviet Union to purchase six million tons of grain per year and the United States to permit the Soviet Union to purchase at least eight million tons per year. If we accept Wolf's characterization of Soviet trade behavior then the Soviet offer curve in Figure 3.7 is changed from ABC to ABD, reflecting the requirement to purchase at least six million tons. The

88

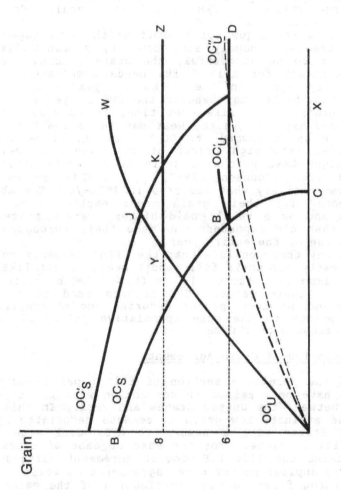

Figure 3.7 The Soviet-American Grain Agreement with Fixed
Consumption Proportions in the Soviet Union

U.S. offer curve may now be OYZ if the United States applies the eight million ton ceiling, as was the case under the grain embargo of 1980, or OYW if the ceiling is lifted. Note however, that the Soviet Union can expect to trade along OYW only at the cost of revealing their own offer curve. The exercise of market power by the United States is now even more attractive than under Wolf's original scenario. If the United States has any market power, then a shift in its offer curve from OC_U' to OC_U'' through the imposition of export restrictions will improve U.S. welfare. While an export tariff is, unfortunately, unconstitutional, the high proportion of wheat to corn in U.S. exports to the Soviet Union and the requirement for the use of U.S. ships for transporting the grain may be seen as shifts in the terms of trade against the Soviet Union. A second advantage of the agreement is that if the Soviet offer curve shifts outward to OC_S', say as the result of a poor harvest, it is now the U.S. offer curve which is unknown -- since the United States has the option of trading at either OYW or OYZ while the Soviet offer curve must be, at least locally, revealed. The only situation which yields some potential benefit for the Soviet Union from the exercise of market power is if trade takes place along the U.S. offer curve along the YZ segment. Here the Soviets would benefit by shifting their offer curve so as to trade at Y rather than at, say, K.

Batra's conclusions regarding Soviet trade behavior are also modified by the provisions of the grain agreement. In Figure 3.8 we depict the Soviet offer curve as OC_S. The U.S. offer curve, OC_U is taken over from Figure 3.7. Under the conditions depicted in this diagram, trade would take place at A if no impediments to trade are imposed by either party. The United States would benefit by restricting imports to eight million tons and trading at B. The Soviet Union, following Batra's optimum tariff strategy would wish to offer the combination C, thus maximizing Soviet welfare. Such an offer would not be very meaningful, however, once the true Soviet offer curve was revealed through negotiation to exceed eight million tons of exports at given prices. Thus, the Soviet Union is only likely to benefit in this situation if it restricts itself to eight million tons, and shifts its offer curve so as to trade at C.

However, unlike Batra's scenario, where retaliation in the form of a rightward shift in the U.S. offer curve may not improve the terms of trade in favor of grain, under the terms of the grain agreement, an opportunity to obtain a greater share of the gains from trade by the United States does exist. By shifting the U.S. offer curve to OC_U' in Figure 3.9, the United States is able to guarantee itself terms of trade TT', and no retalia-

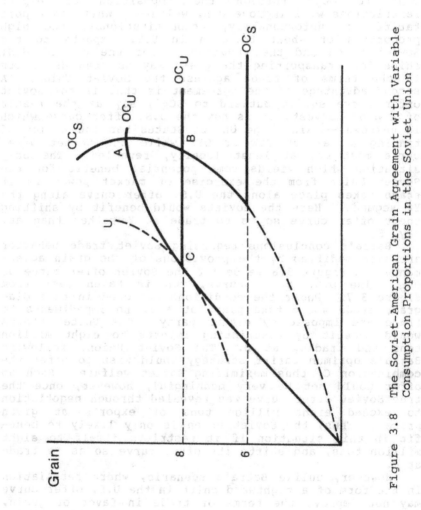

Figure 3.8   The Soviet-American Grain Agreement with Variable Consumption Proportions in the Soviet Union

91

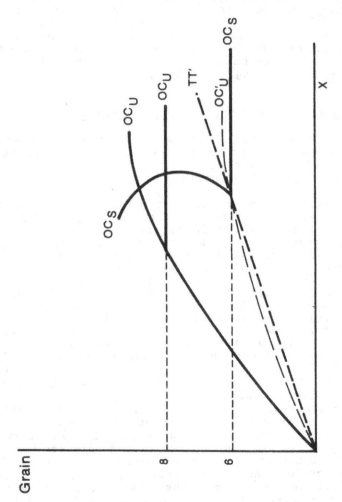

Figure 3.9 Optimal Trade Strategy for the United States When
Soviet Consumption Proportions Are Variable

tory shift in the Soviet offer curve is possible. Thus while Batra's analysis does lead to some possibility for the exploitation of Soviet market power, under the terms of the grain agreement much of this potential is effectively eliminated.

In sum, the provisions of the U.S.-Soviet and thus other similar agreements, appear to serve the interests of the United States in that they protect the United States from the exercise of market power by the CPE and indeed may open options for the exercise of market power by the United States to the detriment of those countries. What the effect of this bilateral agreement, or of similar agreements negotiated by other grain exporting countries, is on the world grain market is the subject of the following section.

Grain Agreements and the World Grain Market

The effect of grain agreements of the type discussed in the preceding section on the world grain market can be analyzed by means of the same offer curve analysis that was used for the analysis of the U.S.-Soviet agreement.

In Figure 3.10a we have the offer curve of the CPEs based on Wolf's assumptions and in Figure 3.10b the offer curve of other grain importing countries. Figure 3.10c represents the sum of the quantities of grain demanded and exports offered in return by all grain importing countries. Thus at terms of trade T_T the CPEs will wish to import OA of grain and the other grain importing countries OB (Figure 3.10b) which is equal to AB in Figure 3.10c.

The signing of a grain agreement between any number of exporters and CPE importers will change the exporters' and CPE importers' offer curves in exactly the same way as described in the foregoing section. If, as in Figure 3.11, the minimum that the CPEs are required to import occurs at terms of trade at which the other grain importers would not wish to purchase any grain, then the grain agreement has no effect on the quantities imported by these countries or on the prices that they pay for grain. The importing countries offer curve is given by ABC. At any grain prices higher than TT_O only the countries that have signed agreements would purchase grain. At lower prices, each group of countries will import exactly the same amounts they would without the agreement unless the CPEs are limited by the exporting countries. Another possibility is that the minimum purchases required of the CPEs would take place at prices at which the other grain importing countries would also wish to import grain. The situation then becomes somewhat more complex. In Figure 3.12 ABC is the CPEs offer curve, and DE the additional demand of

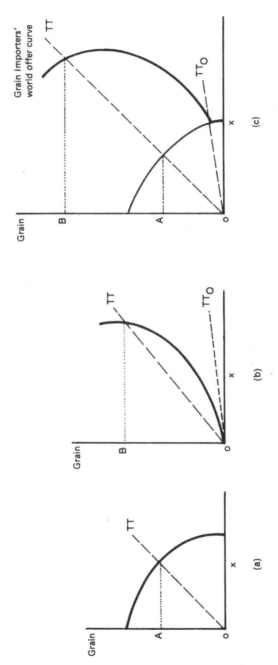

Figure 3.10 Aggregate World Demand for Grain in the Presence of Grain Agreements

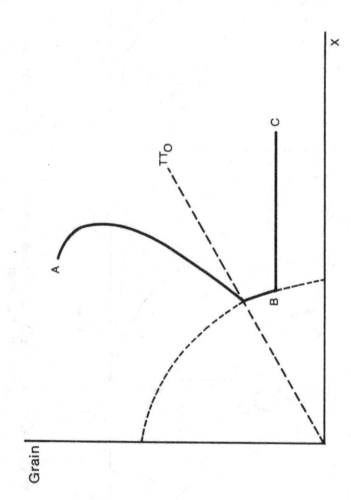

Figure 3.11   Unchanged World Trade in Grains in the Presence
of Grain Agreements

95

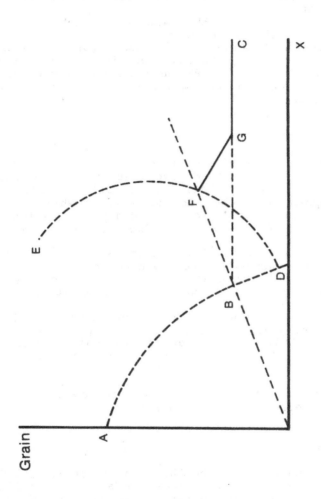

Figure 3.12   Benefits to Grain Exporters from the Existence
of Grain Agreements

the other importers. Because at certain terms of trade the CPEs must now purchase more grain than they would have otherwise, the price and quantity of grain demanded will be higher than otherwise. Thus in the aggregate at some low levels of demand for grain the agreement benefits the exporting countries. Note, however, that if the exporting countries' offer curve intersects EFGC, the importing countries' offer curve, in the EF segment then the existence of the grain agreements has no effect on prices and quantities.

Thus it is evident that whether or not other grain importers are likely to be affected by the signing of grain agreements between CPEs and exporting countries depends critically on the relationship between the minimum quantities that the CPEs undertake to import and the prices at which non-CPE importers would choose not to import grain.

NOTES

*/ The research underlying this paper was supported by a grant from the Deutscher Akademischer Austausch dienst, and National Council for Soviet and East European Research. It was written while I was a guest of the Osteuropa-Institut Munchen to which I am most grateful for a most productive and stimulating environment.

1. Yields, rather than output, were used to offset the effect of changes in area sown to individual crops.

2. Moreover cattle and poultry yield products other than meat, thus also arguing for some moderation in changes in their numbers.

3. This section of the paper draws heavily on another paper published by the author (Brada 1983).

4. See Wolf (1982b) for statistical evidence that the Soviet offer curve does indeed have the shape depicted in Figure 3.4. Moreover, the evidence on East European planners' behavior in the 1970s, presented above, suggests that a similar conclusion can be drawn regarding their offer curve as well.

REFERENCES

Batra, Raveendra N. "The Theory of International Trade with an International Cartel or a Centrally Planned Economy". Southern Economic Journal Vol. 42, No. 3 (January 1976): 364-376.

Brada, Josef C. "The Soviet-American Grain Agreement and the National Interest". American Journal of Agricultural Economics (November 1983): 651-656.

Goldman, Marshal I. Detente and Dollars. New York: Basic Books (1975).

Johnson, D. Gale. The Soviet Impact on World Grain Trade. New York: British-North American Committee (1977).

Vernon, Raymond. "The Fragile Foundations of East-West Trade". Foreign Affairs, Vol. 57, No. 5 (Summer 1979): 1035-1051.

Wolf, Thomas A. "The Theory of International Trade with an International Cartel or a Centrally Planned Economy: Comment". Southern Economic Journal. Vol. 44, No. 4 (April 1978): 987-991.

Wolf, Thomas A. "Optimal Foreign Trade for the Price-Insensitive Soviet-Type Economy". Journal of Comparative Economics. Vol. 6, No. 1 (March 1982a): 37-54.

Wolf, Thomas A. "Soviet Market Power and Pricing Behaviour in Western Export Markets". Soviet Studies. Vol. 34, No. 4 (1982b): 592-596.

4

Import Response, Foreign Exchange Allocation and Inconvertibility in the Centrally Planned Economies

James R. Jones, Hassan Mohammadi,
C. S. Kim, and Joel R. Hamilton

INTRODUCTION

During the decade of the 1970s several forces con-
verged that placed onerous demands on the foreign
exchange purchasing capacities of the centrally planned
economies (CPEs).[1] The explosion of energy prices
worked against the terms of trade of the energy deficit
countries in Eastern Europe. Ambitious capital and
technology import programs were undertaken in anticipation of offsetting the outflow of hard currency with
increased manufactured exports by the end of the
decade. Unfortunately, these programs were poorly
instrumented. Quality and capacity problems continued
to plague industrial exports, plus a worldwide recession
blunted demand further. Ambitious programs to expand
meat production and consumption were launched just when
a series of crop failures and rapidly increasing grain
prices required large outlays of foreign exchange to
sustain these programs.

The culmination of all of this was that these economies were faced with massive deficits in their foreign
exchange receipts relative to outlays. To cope with
these external financial deficits it was necessary to
increase borrowing in international capital markets (see
Chapter 1). Longer term fundamental adjustments were
not undertaken by the authorities until the end of the
1970s and beginning of the 1980s after they had begun to
exhaust their ability to sustain imports via hard currency borrowing.

Typically two-thirds of U.S. sales to the Soviet
bloc centrally planned economies have been comprised of
grain exports. There is little understanding of how
these sales are affected by interactions between variations in world prices and the hard currency position of
these economies. This paper first discusses the incon-

sistency of findings in empirical studies to date in providing reliable elasticity estimates. The second part of the paper looks at conceptual issues that will hopefully aid understanding of how agricultural import decisions in centrally planned economies are affected by variations in world prices, domestic economic considerations and the countries' foreign accounts position as determined by export and import relations and credit access in international financial markets. General equilibrium and partial equilibrium analytical techniques will be adapted to the task of analyzing import response in centrally planned economies. The framework is modified to allow disequilibrium and rationing of commodities to be explicitly treated as one of the issues that arise in the specification and estimation of empirical models of commodity specific foreign trade behavior in centrally planned economies. The response of commodity imports to the effects of domestic economic occurrences and foreign sector phenomena are examined in a simple framework that treats these phenomena as exogenous shifts.

The discussion is limited in scope and ignores or treats very lightly many of the issues important to understanding import behavior in CPEs. It only begins consideration of policy trade-offs facing the planners. From a modeling perspective little is said about whether it is permissible to treat certain shocks or shifts as exogenous, or whether and how the appropriate specification should treat certain variables in an endogenous framework. The paper differs in scope from papers by Portes (1979), Wolf (1980) and others who have considered adjustment in centrally planned economies to external disturbances in that, while they focused upon the macroeconomic aspects of adjustment, the topic here is micro level adjustments of specific agricultural imports to internal and external shocks. Nevertheless it bears some resemblance and debt to these studies to the extent that macro balance-of-payments adjustment processes interplay with specific or micro import considerations.

EMPIRICAL MEASUREMENT OF IMPORT RESPONSE

Conventional wisdom has held generally that import demand elasticity in the case of Soviet type centrally planned economies with respect to variations in world prices approaches zero (e.g. see Brown and Neuberger, et al. 1968; Holzman 1976). It has been argued (Abbott 1979; Bredahl et al. 1979) that agricultural imports from the West are price inelastic. Wolf (1982) explains the case for trade insensitivity in terms of price variations in world markets on the basis that production plans are unlikely to be altered by changes in terms of

trade on the world market. Also planners' welfare functions will reflect an unwillingness to substitute goods
in consumption once consumption targets are set. Commodity inconvertibility and currency inconvertibility
have been cited as insulating domestic prices from world
prices, therefore resulting in world price variations
not being transmitted to domestic prices. The linkage
between domestic prices and world prices is argued to be
effectively severed by use of price equalization subsidies or taxes (Wolf 1980).

The natural inclination is to ask, does the empirical evidence support the inelastic import demand hypothesis? It is difficult at this stage to conclude
where the weight of the evidence lies. Empirical work
applied to agricultural imports in the centrally planned
economies to date has tended to yield wide variations
regarding price response (Desai 1981; Ryan and Houck
1976; Gadur 1981; Jones and Morrison 1976; Abbott 1979;
Konandreas et al., 1984; Mitchel et al., 1984; Carson et
al., 1984; Crane and Kohler, 1984). Table 4.1 which
presents these results indicates that wide variations
occur not only among countries, but across different
studies for the same country or country grouping.

One has to interpret econometric results of import
response with extreme caution. Parameter estimates have
proven to be highly sensitive to the chosen specification even in the case where market type economies have
been studied. Complicating matters in the centrally
planned economies is the deficiency in data to support
the effort. Paucity of data is a less serious problem
now than a decade ago, but still precludes specification
in complete accordance with an ideal theoretical framework. Information regarding domestic prices paid by end
users is needed to determine how domestic prices
influence demand but these data are not consistently
available for all years and countries. Time series data
regarding prices received by producers for feedgrains
and prices paid by producers for inputs are obtainable
in selected cases from official country sources and from
other miscellaneous sources (e.g. see United Nations
Economic Commission for Europe). To the authors' knowledge this covers about half of the CPEs, but does not
include the Soviet Union. Also grain stock data are not
available for most years and countries. Data relating
to foreign exchange reserves, international borrowing,
and international terms of trade are either not consistently available or are based upon estimates.

The task of closing the data gaps alluded to here is
one of the areas that will need much attention before
evidence can conclusively be brought to bear on how to
characterize agricultural import behavior in the centrally planned economies.

Another shortcoming of empirical studies on agricultural import behavior is that they have been designed

Table 4.1 Import Demand Price Elasticity Estimates for Grains and Soybean Meal in the Centrally Planned Economies (values in parenthesis are the t values associated with price coefficient estimates).

Country	E_{gc}^a	E_{hr}^b	E_a^c	E_d^d	E_{oc}^e	E_{msu}^f	E_{jm}^g	E_k^h	E_{ck}^i
Soviet Union	-1.50 (-1.30)	---	0.11* 0.02*	0.00	-2.12 (-2.79)	---	---	---	-1.17
Poland	2.14* (2.40)	0.00	---	---	---	---	0.00	---	---
German Democratic Republic	-0.27 (-0.44)	-0.76	---	---	---	---	0.00	---	---
Czecho- slovakia	0.29* (0.26)	0.00	---	---	---	---	0.00	---	---
Romania	-3.72 (-1.16)	---	---	---	---	---	---	---	---
Bulgaria	-1.02 (-0.54)	---	---	---	---	---	---	---	---
Hungary	---	-7.00	---	---	---	---	0.00	---	---
Yugoslavia	10.71* (1.63)	---	---	---	---	---	0.00	---	---
Eastern Europe	---	---	---	---	---	---	---	---	-1.28
Peoples' Republic of China	-2.57 (-1.80)	---	---	---	---	---	---	---	---
Centrally Planned Economies	---	---	---	---	---	-1.47	---	-34.07	---

NOTES: Table 4.1

*economically perverse signs.

aE_{gc} is corn import elasticity of demand with respect to a change in the price of corn relative to barley estimated by Gadur (1981).

bE_{ht} is based upon price coefficients and data estimated and reported by Ryan and Houck for feedgrain imports in Eastern Europe (1976).

cE_d is import elasticity of feedgrains in the upper number and wheat in the lower number reported by Abbott (1979).

dE_d is the elasticity of grain import demand with respect to implicit price as calculated from a study reported by Desai (1981).

eE_{oc} is the elasticity of demand for Soviet grain imports estimated by Carson, Love, and Griesmar (1984).

fE_{msu} is the elasticity of coarse grain imports demand estimated and reported by Mitchell (Schmitz, et. al., 1981).

gE_{jm} is the elasticity of import demand for soybean meal with respect to price reported in Jones and Morison (1976).

hE_k is the elasticity of United States' export demand to the Soviet Union and Eastern Europe estimated by Konandreas, Bushnell, and Green (1978).

iE_{ck} is the elasticity of food import demand with respect to a change in price reported by Crane and Kohler (1984).

and interpreted in an ad hoc fashion rather than on the basis of a well-developed conceptual framework. A common characteristic of the previous works is that they were specified within a framework that inadequately accounted for the interrelationships between imports, world prices, and foreign exchange availability. The simplest way to incorporate periods of foreign exchange rationing into import demand functions is to employ dummy variables for periods when such conditions prevail. It was pointed out in Chapter 1 that the general conditions of inconvertibility and foreign exchange shortages have been long-standing in the Soviet bloc centrally planned economies. The question is not just when, but how have such conditions affected specific commodity imports, so the dummy variable technique is limited for this purpose.

Use of an import capacity variable as a shifter of the import demand function is a relatively common method of allowing for situations where inconvertibility and foreign exchange shortages require explicit administratively determined allocation of foreign exchange to be undertaken by the government authorities in the process of importing (Scobie 1981). Frequently foreign exchange earnings from exports, and/or international reserves are used as import capacity variables that shift the import demand function. This is still a very crude way to account for import capacity. In a short run analysis it is necessary to supplement these two capacity variables with credit and expenditures for other imports in an attempt to more fully capture the effect of explicit foreign exchange allocation decisions on agricultural imports in the case of the Soviet bloc centrally planned economies. It has been noted by Scobie that import studies generally fail to give explicit recognition to competing uses of foreign exchange and to providing an underlying behavioral model of the balance-of-payments adjustment process. The models reported above fail to give explicit recognition to changes in borrowing and foreign exchange allocations to other import expenditures. The relationship between import response, price variation and foreign exchange allocation is developed theoretically in the context of the centrally planned economies below. It will be argued that the hard currency effect of world price movements needs to be taken into account and that in this context a tendency to assume that imports in centrally planned economies are perfectly price inelastic can be extreme.

THE THEORY OF IMPORT BEHAVIOR IN A CPE

Analytical approaches applied to market economies may be applied if modified properly to behavior in centrally planned economies (Batra 1976). Figure 4.1 pre-

105

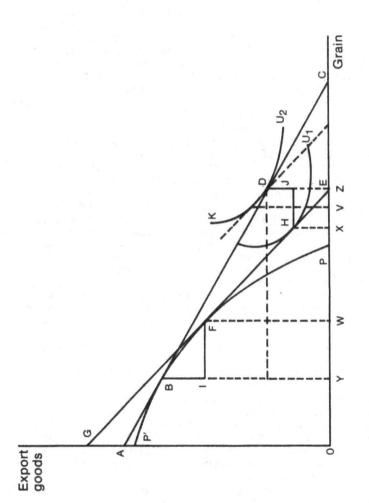

Figure 4.1 Grain Import Response to Price: Market Economy Case

sents a general equilibrium analytical framework which can be used to show the effects of world price changes on imports. In a market economy with resources and consumption free to respond to domestic and international price signals, an increase in the price of imports will have the effect shown in Figure 4.1. Before an increase in the world price of grain the country would import an amount of grain equal to YZ. After an increase in price, which pivots the price line to GE from AC, imports fall to WX. The response of domestic production to a higher world price for grain obtained on the international market place is IF(YW). This added production substitutes for imports. Grain imports are further reduced by HJ(XZ) due to a curtailment in consumption. If the loss in real income were compensated for, this consumption effect could be decomposed into a consumption substitution effect VZ and a hard currency effect, XV (approximately analogous to an income effect in terms of international purchasing power). This general equilibrium formulation has been used after proper modification to analyze trade behavior in centrally planned economies by Batra (1976) and Wolf (1982).

Wolf adapted the conventional general equilibrium model above to reflect the inflexibility that may exist on behalf of planners towards altering consumption and production targets in the short run. Brada uses the same framework in this book (Chapter 3) to analyze the arguments for long-term bilateral grain agreements with the Soviet Union. Their adapted model is reproduced in Figure 4.2. Production possibilities are reflected in q'q. Authorities are assumed to be unwilling to substitute exports (represented on the vertical axis) for imports of grain and this is reflected by the L shape of their indifference curves U_1 and U_2. Moreover as the terms of trade TT vary, production remains fixed at P.

Wolf and Brada's assumptions of a fixed point production function and a fixed-coefficient planner's welfare function suggest that import response to changing prices is less than would be the case in market type economies. However, price and substitution insensitivity within the domestic economy does not preclude import variations in response to changes in world prices or terms of trade. Note in Figure 4.2 for example that pivoting the terms of trade line under the Wolf/Brada assumptions still causes imports to vary. Even if substitution of consumption and production does not occur, the income or hard currency effect of a price reduction of grain from TT to TT' allows imports to increase from M_G to M_G'. At the improved terms of trade, planners can move from U_1 to a higher level of satisfaction U_2 by importing more grain. Importers would logically respond in some degree to world prices regardless of the degree of convexity in their utility functions and the

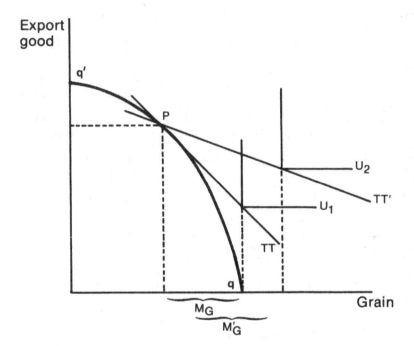

Figure 4.2 Grain Import Response to Price: Centrally
          Planned Economy Case

fixity of domestic production because of the hard currency effect of world price movements.

The above illustration suggests that planning authorities could logically be expected to respond to world price signals on imports, even if they are inflexible on production and consumption substitution possibilities, but it ignores other options also available to the authorities. Figures 4.1 and 4.2 are based upon a balanced trade identity. Following Desai and Bhagwati (1981), the general equilibrium framework can be adapted to a situation where balance of trade current account deficits are absorbed by borrowing in international capital markets. In Figure 4.3 we portray an open account deficit offsetting the hard currency effect by distinguishing between a balanced expenditure terms of trade line DA and a parallel international expenditure line BE with consumption of export goods and import goods occurring at C as before an increase in grain prices.

A diagrammatic apparatus (see Figure 4.4) analogous to the excess demand partial equilibrium framework commonly used to examine import behavior in market type economies is modified to reflect the import decision process in centrally planned economies. This approach will allow the focus to be placed on the relationship between domestic price insulation, world price movements, foreign exchange allocation and import behavior. Prices on the vertical axes are expressed in terms of the foreign hard currency equivalent (dollars) of the home price (based on the official external rate of exchange) and the foreign currency import price.[2] Goods imported are not simply a reflection of excess demands at world market prices. Currency inconvertibility reigns, and thus foreign exchange must be allocated to importers by the Ministry of Finance. Furthermore commodity inconvertibility exists in the sense that domestic prices are insulated from world prices (at least in the short run) principally through the mechanism of a price equalization tax (subsidy). Consequently domestic demand functions (D) and supply functions (S) cannot be viewed in the conventional sense as giving rise to import demand functions. These functions are included in Graph 4.4a for purposes of discussing the comparison or contrast to market economies and to recognize that these functions may affect import decisions at least indirectly but not automatically.

Domestic consumer prices and producer prices are set by the state. Because agricultural prices are often fixed for rather long periods of time (see Chapter 1) domestic prices do not usually reflect movements in international prices in the short run. However it was observed in Chapter 1 that in the late 1970s and early 1980s, the authorities adjusted prices because of their acute balance-of-payments problems. This could be

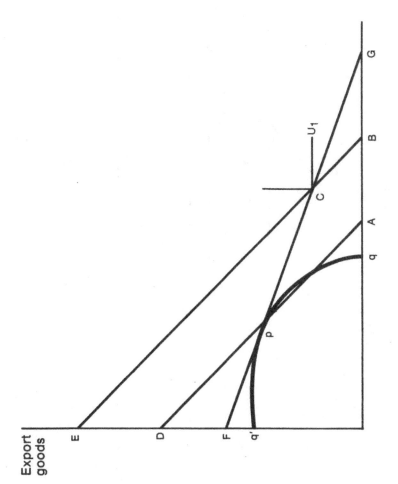

Figure 4.3 Response to Price and Credit Offset

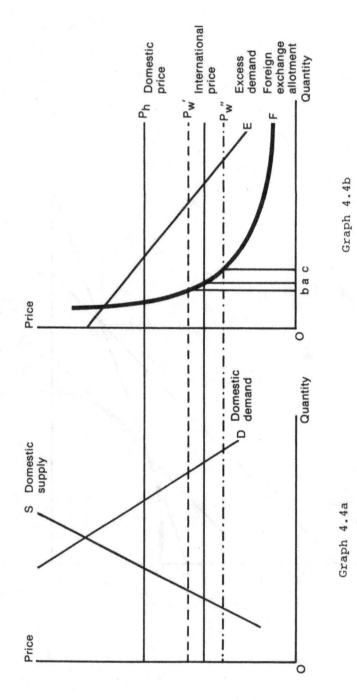

Graph 4.4a

Graph 4.4b

Figure 4.4 Hard Currency Constraint and World Price Variation

viewed as a lagged adjustment to world prices since a portion of the balance-of-payments pressures were due to the rising prices of commodities earlier in the 1970s. What can be said is that prices are infrequently altered but when adjustments are made they may be substantial. Also the exchange rate is infrequently altered except in the case of Yugoslavia and more recently Poland and Hungary. The valuta prices of imports (the foreign currency price times the official exchange rate) seldom equals the home price. Even if official exchange rates are varied, which is seldom, they usually have no effect on domestic or foreign prices (Wolf 1980, p. 95). Consequently the CIF world price of the commodity (P_w) is shown to be separated from domestic prices with the difference constituting a price equalization tax (subsidy). It is common to observe contrived gaps between producer and consumer prices also, but we will ignore this until later. Inland transportation costs, foreign trade organization commission charges, storage, and other marketing charges are also ignored for the sake of simplicity. This convenient assumption allows us to speak of a common home price (P_h).

In the conceptual framework, it is explicitly recognized that currencies are inconvertible for international transactions. In Graph 4.4b, a rectangular hyperbola (F) depicts a given foreign currency allocation provided to the foreign trade organization that purchases imports. When convertibility prevails, imports need not confront a predetermined hard currency constraint since foreign exchange can be purchased with domestic currency. However, in the centrally planned economies, inconvertibility results in a situation where end-users or the specific foreign trade organization (FTO) that transacts foreign purchases must be allotted foreign exchange.[3] It is this allotment that sets the position of F. The curve takes the form of a rectangular hyperbola because the total value of imports that can be acquired from this allotment is constant. The foreign currency allocation is assumed to be authorized by a higher level organization such as the state planning commission, the ministry of finance, or the foreign trade bank.[4]

A notable characteristic of the above conceptual framework is that actual or "effective" import demand at the FTO level is a function directly of the foreign currency allocation rather than the difference between domestic supply and demand at any given price whenever the hard currency allocation constraint is less than excess demand (E) in the relevant price range.[5] If for the moment we ignore wedges between import, producer, and end-user prices, the curve E is unconstrained or nominal excess demand. Given official exchange rates and domestic purchasing power, end-users would be willing to buy the quantities of imports lying along the

notional excess demand function. However actual imports
are constrained by the state monopoly buyer's appropri-
ated level of available hard currency along F, which
defines the "effective" or constrained level of demand.
Letting M represent the physical volume of demand for
imports, as above, P_W the international price, F a
given allocation of foreign exchange, and ignoring other
non-price determinants of import demand, the import
relation can be written as:

$$(1) \quad M = \begin{cases} \alpha - \delta + P_W \, (\beta - \epsilon) & \text{if } F \geq P_W \cdot E \\ \\ \dfrac{F}{P_W} & \text{if } F < P_W \cdot E \end{cases}$$

where α and δ are intercepts, and $\beta < 0$ and $\epsilon > 0$ are
price coefficients of domestic demand and supply rela-
tions, respectively.

For estimation and analytical purposes this specification has two limitations. It is awkward due to the
possibility that different import observations may fall
on different functions. That is demand may not always
be constrained by foreign exchange availability. While
the inclination of some analysts is to suspect that a
hard currency constraint always prevails this is an
empirical issue. More importantly, it does not take
into account the possibility that the authorities'
allocation of foreign exchange may be responsive in some
degree to world price movements. The interaction
between imports of the commodity in question and world
prices and adjustments in foreign exchange uses and
sources will be addressed later. This above conceptual
framework has obvious limitations but it is useful in
demonstrating the implications of world price movements
on imports in the case where foreign exchange is predetermined at a given level.

Given the world price (P_W) and the foreign currency allotment (F), imports under the above assumptions
would obtain at oa in Graph 4.4b. Now we return to the
argument that because the foreign price is divorced from
domestic price, movements in world prices will not cause
changes in imports by centrally planned economies.
Investigation of Graph 4.4b shows that this will not be
the case if a binding "hard currency constraint" will
not allow the desired level of end use purchases.
Upward price movements to P_W' force curtailment of
imports and downward movements, (i.e., to P_W'') permit
expansion of imports even though the domestic price paid
by end-users is perhaps the same as before. Indeed, if
these conditions prevail and the foreign exchange allo-

cation is constant, the elasticity of import demand would be unitary since the total value of import expenditures would be constant regardless of the level of price.

Bredahl, Meyers, and Collins (1979), following the formulation of export (import) demand suggested by Tweeten (1967), developed the important theoretical and empirical role that price transmission elasticity plays in shaping the elasticity of demand for U.S. agricultural products in world markets. In this formulation the elasticity of export demand (E_{ef}) can be expressed as:

$$(2) \quad E_{ef} = \Sigma_i E_{pi} \, E_{edi} \cdot \frac{Q_{mi}}{Q_{ef}} - \Sigma_j E_{pj} \cdot E_{esj} \cdot \frac{Q_{xj}}{Q_{ef}}$$

where E_{pi}, E_{edi}, and Q_{mi} are the price transmission elasticity of domestic price with respect to world price, excess demand elasticity, and volume of imports respectively for each importing country or region i. The price transmission elasticity, E_{pj}, and excess supply elasticity E_{esj} reflect the response of competing exporters to world price movements. The total volume of trade or imports is Q_{ef} and Q_{xj} is the volume of exports from competing country j.

Using this formulation, if domestic prices are insulated completely from world price movements ($E_{pi} = 0$), Bredahl, et. al. argue that the response of importers to a change in world price will be zero, or import elasticity with respect to world price will be perfectly inelastic.

In the case of Figure 4.4 there may be a grouping of importers for which a binding hard currency constraint overrules this. Breaking importers into a category with unconstrained imports (group a) and another category (category b) operating under a binding hard currency constraint, the expression relating to the impact of importers response in equation 2 can be rewritten as:

$$(3) \quad \Sigma_i E_{pi} \, E_{edi} \frac{Q_{mi}}{Q_{ef}} = \Sigma_a E_{pa} \cdot E_{eda} \cdot \frac{Q_{ma}}{Q_{ef}} + \Sigma E_{pb} \cdot E_{edb} \frac{Q_{mb}}{Q_{ef}}$$

If a binding and constant hard currency constraint is assumed $E_{pb} \cdot E_{edp} = -1$ and this obviously has a bearing on total demand elasticity.

Effectively what occurs is that international price variations are directly transmitted to imports in hard currency constrained economies in the b grouping even though domestic prices do not vary, i.e. $E_{pb} = 0$. Queues replace price signals as the devices that ration imports among consumers or end users. If we accept Holzman and others' contention that centrally planned economies are predisposed to incur balance-of-payments

difficulties and are therefore required to ration
foreign exchange, import response to world price movements may occur in spite of a tendency to insulate
domestic prices from world prices. A critical issue is
what governs the decision to allocate foreign exchange.

FOREIGN EXCHANGE ALLOCATION RESPONSE

Focusing on the foreign exchange allocation function
in situations where it represents the effective import
demand function clarifies the relationship between
import and world price movements. If the foreign
exchange allocation is shifted to the right (Figure 4.5)
in response to higher world prices in order to hold the
physical volume of imports constant, import demand would
be perfectly inelastic. The response to world price
movements depends on whether foreign currency allocations are really binding constraints or to what extent
authorities are willing and able to vary the foreign
exchange allotment in response to world price changes.

An array of options available to the authorities in
the centrally planned economies can be seen in Figure
4.5. Assume initially that the price paid by end users,
P_C, prevails along with a price, P_p, paid to producers (the difference between P_p and P_C is covered
by state subsidies) and that the hard currency appropriation is set at F. At prices P_C and P_p, a deficit
in domestic availability ab = oe would exist. At the
world price P_W, imports equal to of would leave an
unfulfilled demand equal to fe.

The link between domestic adjustments and import
decisions, of course, is the appropriation of hard currency. If the world price increases to P_W' and no
corresponding action is taken to increase the foreign
exchange allocation, the amount of imports purchased
would fall to og increasing the shortfall to be rationed
from fe to ge. The authorities have several options:
1) queues can be allowed to increase, or if the authorities decide to avoid increased nonprice rationing; 2)
the hard currency appropriation could be increased shifting F upward and to the right to F', 3) prices paid by
end users could be raised, reducing demand pressure; 4)
producers' prices could be raised, increasing the quantity supplied domestically; 5) inputs could be made
available in greater quantities shifting domestic
supply; or 6) a combination of the above courses of
action could be taken.

The discussion to this point suggests that import
behavior in the centrally planned economies cannot be
described by the simple interaction between market
forces nor at the other extreme by the extreme assumption that no response will occur. Actually a complex

115

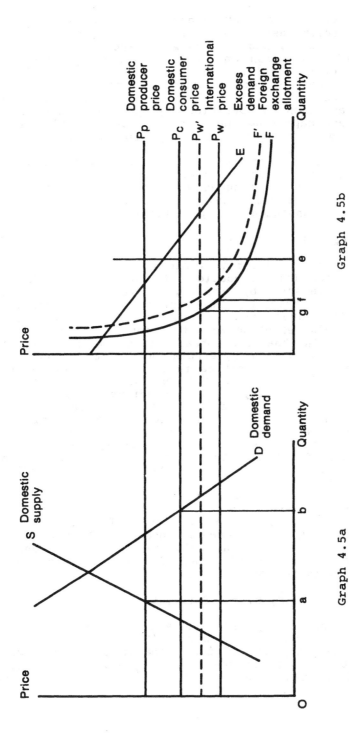

Graph 4.5a

Graph 4.5b

Figure 4.5 Foreign Exchange Allocation Response

array of options are available to the planners. The ultimate relationship between world price movements, domestic policies, and imports will depend on the explicit official decision on how much foreign exchange to allocate toward purchases of the commodity in question. Below the concept of a foreign exchange allocation response (FEAR) function is developed and offered as one possible approach to operationally specifying this relationship and capturing the underlying behavioral characteristics of the import decision process in centrally planned economies.

The allocation of foreign exchange expended for a given commodity is by definition equal to the value of imports. If it is assumed that all foreign exchange allocated to the importing agency is expended and no imports are transacted on a pure barter basis, then [6]

$$(4) \quad F = P_W \cdot M$$

which permits an import response equation to be specified in which the allocation of foreign exchange can directly be measured by treating the value of imports as the dependent variable. This approach has the merit of allowing the investigator to directly focus on allocation of foreign exchange as a function of selected shift variables. In the conceptual framework referred to above, the rectangular hyperbola may shift depending on if, and how, the authorities respond to domestic supply and demand factors and foreign sector developments affecting availability and expenditures of foreign exchange. Following this approach, the concept of a foreign exchange allocation response (FEAR) function is introduced here to express the relationship between the value of imports for which foreign exchange is allocated by the authorities and the arguments or factors that determine this allocation. In other words, it provides a way of looking at the various factors that determine the actual allocation of foreign exchange F and indirectly the level of imports. One can postulate that foreign exchange allocation may be a function of the world price (P_W) and domestic demand (D) and supply (S) considerations and the balance-of-payments standing of the economy (B) in which case the FEAR function could be expressed in the general form

$$(5) \quad F = F (P_W, D, S, B) = F (P_W, E, B)$$

if demand and supply shifters are treated jointly as they affect notional excess demand (E).

The elasticity of foreign exchange allocation with respect to world price movements ($E_{f,Pw}$) can be shown to be unitary plus the partial import price elasticity ($E_{m,Pw}$):[7]

(6) $E_{f,Pw} = 1 + E_{m,Pw}$

or

(6)' $\dfrac{\partial F}{\partial P_W} \cdot \dfrac{P_W}{F} = 1 + \dfrac{\partial M}{\partial P_W} \cdot \dfrac{P_W}{M}$

The coeffieicnt $E_{f,Pw}$ may supercede the price transmission elasticity E_{pi} alluded to in equation 3. World price movements are transmitted directly to import response as we see when we conveniently rearrange equation 6 in terms of the price elasticity of import demand:[8]

(7) $E_{m,Pw} = E_{f,Pw} - 1.$

In this expression as in the graphical presentation (Figure 4.4) it becomes apparent that in cases where the allocation of foreign exchange is not changed in response to a change in the world price ($E_{f,Pw} = 0$) the elasticity of import demand with respect to a change in world price would be unitary. Alternatively if world price variations are proportionately compensated for by foreign exchange allocations ($E_{f,Pw}=1$) then import demand is perfectly inelastic.

Estimating a FEAR function and deriving the implicit import elasticity has the advantage of focusing attention on planner's explicit policy decisions regarding foreign exchange allocation. A recent Rand study (Crane and Kohler, 1984) in essence followed this route. Their researchers employed a linear expenditure model adopted from consumer demand theory to derive expenditure elasticities for food, raw materials, machinery, and other manufactured products. In general the Rand elasticity estimates indicated that "price changes have little bearing on the amount of hard currency spent on a particular commodity by the Soviet Bloc, own price elasticities were close to negative one, [and] cross price elasticities were not significantly different from zero" (page vi).[9]

Perhaps the most noteworthy feature of this particular study is that it focuses on the critical relationship between foreign exchange allocations to import food products as opposed to other categories of imports. To the apparent surprise of the authors, expenditure elasticities for food and raw materials suggested slightly more import response to price and hard currency variations than was the case for manufacturers. Whether this finding will stand up to further investigation or not it at least lends support to the need to specify the FEAR function in a way that considers the interaction between allocations of other imports and the good in question.

118

SUMMARY

In this chapter we have explored empirical evidence and theoretical issues pertaining to the import elasticity of agricultural imports in the centrally planned economies in view of a tendency to insulate domestic prices from world price movements. We have concluded that in the context of foreign exchange allocation and rationing a direct correspondence between domestic price transmision approaching zero and import elasticity approaching zero breaks down. Ultimately the responsiveness of imports to world price movements will hinge on the foreign exchange allocation response of authorities to a change in world price.

NOTES

1. In this paper, the CPE grouping includes the Soviet Union, the Eastern European economies and China.

2. This official exchange ratio generally does not reflect true relative values of foreign currencies and the home currency since inconvertibility reigns. Domestic and foreign prices are not organically linked; i.e., the exchange rate serves only an accounting function rather than acting as a policy instrument. (Wolf, June, 1985) summarizes the relationship between domestic prices and foreign currency prices (ignoring marketing costs) in CPEs as:

$$P_i = P_i^* e'(1-\alpha)$$

where P_i^* is the foreign currency price of the good in question, e' is the official external exchange rate (for instance, rubles or forints per dollar), P_i is the domestic currency price at which the good is transferred between foreign trade organizations and domestic enterprises, and represents the export variable implicit tax (subsidy) rate on the commodity. In our notation the home price, P_h expressed in foreign currency is:

$$P_h = \frac{P_i}{e'} = P_w(1-\alpha)$$

where the world price P_w is simply the foreign hard currency price, i.e.:

$$P_w = P_i^*.$$

3. In our example, we can arbitrarily think of end users as feed milling and compounding agencies or enterprises as the case may be, although the thought could be extended to livestock producers or even the ultimate consumer of meat products, etc.

4. Cases exist where end-using enterprises may use their own hard currency earnings to finance imports, but no example of this is known to apply to major agricultural imports and this possibility is ignored in this analysis.

5. Of course the extent of the gap between excess domestic demand and actual outlays for imports presumably would have some influence on the authorities' decision to allocate foreign exchange toward purchase of the commodity in question. This relation is addressed later in the paper.

6. Barter and countertrade transactions are common means of financing imports in the centrally planned economies; but, these arrangements have been relatively rare in grain transactions which comprise the bulk of agricultural trade.

7. Taking the differential of the identity

a) $F = M \cdot P_W$

with respect to price gives

b) $\dfrac{\partial F}{\partial P_W} = M \cdot \dfrac{\partial P_W}{\partial P_W} + P_W \cdot \dfrac{\partial M}{\partial P_W} = M + P_W \cdot \dfrac{\partial M}{\partial P_W}$

This relation can be reexpressed in elasticity form as:

c) $E_{f,Pw} = \dfrac{\partial F}{\partial P_W} \cdot \dfrac{P_W}{F} = M \cdot \dfrac{P_W}{F} + P_W \cdot \dfrac{\partial M}{\partial P_W} \dfrac{P_W}{F}$

$= 1 + \dfrac{\partial M}{\partial P_W} \cdot \dfrac{P_W}{M}$

d) $E_{f,Pw} = 1 + E_{m,Pw}$

8. We are indebted to Mr. Shahid Perwaiz for working with us on mathematical manipulation that allowed us to derive the relationships expressed in equations 6 and 7 between the elasticity of foreign exchange transmission and the elasticity of import demand.

9. Corollary but separate work is in process at the University of Idaho in conjunction with the International Economics Division of the U.S. Department of Agriculture.

REFERENCES

Abbott, Philip C. "Modeling International Grain Trade with Government Controlled Markets." Amer. J. of Ag. Econ. 61 (1979): 22-31.

Batra, Raveendra N. "The Theory of International Trade with an International Cartel or a Centrally Planned Economy." Southern Economic Journal 42 (January 1976): 364-376.

Brada, Josef C. "The Soviet-American Grain Agreement and the National Interest." Amer. J. of Ag. Econ. (November 1983): 651-656.

Bredahl, Maury E., William H. Meyers, and Keith J. Collins. "The Elasticity of Foreign Demand for U.S. Agricultural Products: The Importance of the Price Transmission Elasticity." Amer. J. of Ag. Econ. 61 (February 1979): 58-63.

Brown, Alan A. and Egan Neuberger, eds. International Trade and Central Planning. Berkeley: Univ. of Calif. Press (1968).

Carson, Richard T., Alan Lane, and Fabien Keller-Griesmar. The Soviet Grain Import Decision as a Short Term Control Problem. Selected paper presented at the American Agricultural Economics Association Annual Meetings. Cornell University (August 1984).

Central Intelligence Agency National Foreign Assessment Center. Estimating Soviet and East European Hard Currency Debt. ER80-10327 (June 1980): 7, Table 4.

Crane, Keith W. and Daniel F. Kohler. "The Effects of Export Credit Subsidies on Western Exports to the Soviet Bloc." Santa Monica, CA., Rand (A Rand Note prepared for the Office of the Undersecretary of Defense for Policy, N-2106-USPD) June, 1984.

Desai, Padma. Estimates of Soviet Grain Imports in 1985: Alternative Approaches. International Food Policy Research Institute, Research Report No. 22, (February 1981).

Feder, Gershon, and Richard E. Just. "An Analysis of Credit Terms in the Eurodollar Market." European Econ. Rev. 9 (1979): 221-243.

Gadur, Ahmed Ibrahim. "An Economic Analysis of Feed Grain Import Dependence in the Centrally Planned Economies," M.S. Thesis, University of Idaho, Department of Agricultural Economics and Applied Statistics (1981).

Hanson, Philip. "The End of Import Led Growth? Some Observations on Soviet, Polish and Hungarian Experience in the 1970s." Journal of Comparative Economics. 6 (1982): 136.

Holzman, Franklin D. International Trade Under Communism. New York: Basic Books, Inc. (1976).

Holzman, Franklin D. "Some Theories of the Hard Currency Shortages of Centrally Planned Economies." in U.S. Congress, Joint Economic Committee. Soviet Economy in a Time of Change. 2 (1979): 297-316.

Jones, James R. "Inconvertibility, Foreign Exchange Allocation, and Agricultural Import Response in the Soviet Bloc Economies." paper presented at the American Agricultural Economics Association meetings, Logan, Utah (August 1982).

Jones, James R. and W. R. Morrison. Import Demand for Soybeans and Soybean Products in Eastern Europe. Univ. of Arkansas, Ag. Exp. Sta. Bulletin 803, (March 1976).

Konandreas, Panos, Peter Bushnell, and Richard Green. "Estimation of Export Demand Functions for U.S. Wheat." Western Journal of Agricultural Economics. (July 1978): 39-49.

Portez, Richard C. "Internal and External Balance in a Centrally Planned Economy." Journal of Comparative Economics. (1979): 325-345.

Ryan, Mary E. and James P. Houck. Eastern Europe: A Growing Market for U.S. Feed Grains. Univ. of Minn.: Econ. Report 76-7 (May 1976).

Scobie, Grant M. Government Policy and Food Imports: The Case of Wheat in Egypt. International Food Policy Research Institute Research Report 29 (December 1981).

Schmidt, S.C., J.R. Jones, D. M. Conley and A. R. Bunker. Cooperative Grain Opportunities in Eastern Europe. Cooperative Marketing and Purchasing Division, Agricultural Cooperative Service, U.S. Department of Agriculture ACS Research Report No. 21 (May 1984).

Schmitz, Andrew, Alex F. McCalla, Donald O. Mitchell and Colin A. Carter. Grain Export Cartel. Cambridge, Mass.: Ballinger (1981).

United Nations Economic Commission for Europe, Food and Agricultural Organization. Prices of Agricultural Products and Selected Inputs in Europe and North America, various years.

Wolf, Thomas A. "Optimal Foreign Trade for the Price-Insensitive Soviet Type Economy." Journal of Comparative Economics. (1982): 37-54.

_____. "Exchange Rate Systems and Adjustment in Planned Economies." International Monetary Fund Staff Papers. Vol. 32, No. 2, (June, 1985), pp. 211-247.

_____. "On the Adjustment of Centrally Planned Economies." IN East European Integration and East-West Trade ed. by Paul Marer and Michael Montias. Bloomington: Indiana University Press (1980): 86-111.

World Bank. Commodity Trade and Price Trends. (August 1981): 30.

5

The Centrally Planned Countries' Livestock Product and Feed Grain Systems

Kenneth B. Young and Gail L. Cramer

INTRODUCTION

In this chapter we discuss the relative importance of livestock production in centrally planned countries to international trade of grain and oilseeds, the basis for increasing trade dependence of centrally planned countries on western countries, and limitations to further growth in international trade. We then present an overview of: 1) the model used to estimate feed/livestock sector relationships in centrally planned countries, 2) the application of the model to assess current operation of the livestock/feed systems, and 3) the use of the model to project future feed requirements under alternative production and consumption targets in centrally planned economies.

TRADE RELATIONSHIPS WITH THE WEST

Imports of grains and oilseeds to centrally planned countries account for a significant share of total world trade. During the 1982/83 crop year the Soviet Union, Eastern Europe, and the Peoples' Republic of China (PRC) together imported 37 million metric tons (mmt) of wheat and wheat flour, and 18 mmt of coarse grains. This represented about 30 percent of total world grain trade excluding rice (USDA 1983). Oilseed imports to centrally planned regions have constituted about 30 percent of annual total world trade in recent years.

The relative importance of centrally planned countries as grain importers has shifted dramatically over the past two decades. Prior to 1960, Western Europe was the only major grain import market, and centrally planned countries were of minor importance in world trade. However, the situation changed after the early 1960s as imports to centrally planned countries figured prominently in the rapidly increasing export market of the 1970s. The severe shortfall in the 1973 world grain market was attributed in large part to the unprecedented activity of the Soviet Union as a dominant grain buyer. The PRC has also purchased significant volumes although more recently it has reduced its imports. Eastern Europe has been a fairly minor but consistent grain importer since the 1960s.

The growing trade deficit in both grain and oil-seeds in centrally planned countries is attributed to a combination of factors. Although increasing population and production problems have contributed to the growing deficit, the chief factor has been expanded livestock feeding to improve living standards, causing demand to increase faster than production capacity. In contrast to other growing grain import markets, the centrally planned countries as a group have been irregular buyers due to the dominance of the Soviet Union in this group and its past record of making erratic purchases from year to year. The Soviet Union was an especially dis-ruptive force in the early 1970s when the Soviets first began purchasing sizable amounts of grain on an ad hoc basis to make up recurring shortages in domestic produc-tion. Concern over these continuing disruptions led to negotiation of a five-year grain supply agreement between the United States and the Soviet Union to ship eight mmt per year in the period 1976/77 to 1980/81. Another five-year agreement was signed for the period 1983/84 to 1988/89. Minimum shipments are to be nine million tons of wheat, corn, and soybeans.

Aside from the problem of wide variation in grain yields from year to year in the Soviet Union, the cen-trally planned countries as a group have less predict-able demand for grain and oilseed imports than most other importing regions due to the nature of their planned process. In contrast to free market economies, they exercise tight political control over both consump-tion and production of most agricultural products. The Soviet Union, for example, in 1973 reversed prior policy of periodically reducing livestock inventories to adjust livestock production to short crop years and began importing large quantities of grain thereafter to main-tain livestock production in shortfall years. Strict consumer rationing of agricultural products in short supply is a common occurrence in centrally planned coun-tries and can be used effectively to reduce dependence on imports depending on the policy of the central government. Currently, however, the Soviet Union is using a large-scale subsidy program to increase the out-put of agricultural commodities. Substantial subsidies are made on farm commodities as well as purchased farm inputs. Net subsidies to agriculture amounted to 31.4 billion rubles on agricultural products and 3.6 billion for agricultural inputs (Treml 1983). The Soviet Union, in particular, has a longstanding reputation of restric-ting consumer products to concentrate on production of

industrial goods and military equipment. Both Eastern
Europe and the PRC are currently in need of substantial
new capital investment for industrial development to
improve their export capability. This could have a
restrictive effect on future production of nonessential
consumer items and on feed imports for livestock production.

Both the Soviet Union and Eastern Europe have a
wider range of flexibility to influence overall grain
use compared with the PRC because of the heavy emphasis
on grain feeding in these countries. Most of the grain
use in the PRC is for direct human consumption and only
32 mmt, about 20 percent of total use, is allocated for
feed use.[1] On the other hand, the Soviet Union and
Eastern Europe fed 188 mmt in the 1981/82 crop year,
about 35 percent of total world feed use. Feed use comprised 56 percent of total use in the Soviet Union and
67 percent in Eastern Europe. Eastern Europe and the
Soviet Union also use a high proportion of wheat for
feed depending on the size and quality of their wheat
crop. They fed 57 mmt in the 1981/82 crop year, about
69 percent of total world feed use of wheat.

Since the centrally planned countries are currently
deficit in both grain and oilseeds, development of their
livestock industry over the next few years is dependent
on the availability of foreign exchange or access to
credit. Eastern Europe has currently accumulated a
large balance-of-payment deficit and their credit standing is also in some jeopardy due to the recent Poland
crisis and their continuing annual trade deficit. The
Soviet Union, on the other hand, is much less dependent
on credit for financing imports and currently has much
more favorable long-run prospects of earning additonal
foreign exchange such as through the sale of natural gas
to Western Europe. A major effort is underway in the
PRC to expand their export industries. This could
result in a significant boost in foreign exchange earnings which, in turn, would stimulate major growth in the
PRC livestock industry.

Grain and oilseed imports from western countries
are the major trade requirements of the centrally
planned countries. These requirements are closely
intertwined with their domestic livestock production
programs, particularly in the Soviet Union and Eastern
Europe. Although the livestock industry is still
largely undeveloped in the PRC, the Chinese are currently in the process of adapting western technology to
modernize their livestock industry, which could ultimately lead to additional use of imported concentrate
feeds.

The analysis presented here includes a current
assessment of livestock feeding practices in centrally
planned countries and projections of probable import
requirements in the year 2000 to sustain their livestock

programs. Projections for 2000 are based on alternative
growth rates for domestic crop production and national
consumption targets. Simulation techniques are used to
estimate livestock consumption requirements, and an
input-output model is used to project trade requirements
under the alternative growth rates.

DESCRIPTION OF MODEL AND DATA SOURCES

Interrelationships between the feed and livestock
sectors of centrally planned countries are evaluated
with the use of an input-output model. Major components
of the model are: 1) a livestock nutrient requirement
matrix, 2) a feed supply matrix, 3) a livestock offtake
matrix, and 4) a feed/livestock system input-ouput
matrix to evaluate the total flow of feed inputs through
the livestock system. A description of each model component follows.

Nutrient Requirement Matrix

Nutrient requirements for different classes of
livestock are estimated in terms of megacalories of
metabolizable energy (Mcal ME) and kilograms of diges-
tible protein (kg DP) as illustrated in Figure 5.1. Biological submodels are used to estimate the nutrient
requirements of different livestock classes based on
national phenotypic data and general indicators of productivity such as meat offtake relative to inventory.
Productivity indicators were derived from periodic
reports of USDA Foreign Agricultural Service (FAS), Food
and Agricultural Organization (FAO) of the United
Nations and other miscellaneous sources. The biological
model for ruminants is a computer simulation model that
was formerly used at Winrock International to estimate
feed energy requirements and to project outputs of meat,
milk, fiber, and work from ruminant production systems
(Nguyen and Fitzhugh). Biological models for poultry
and swine are relatively simple FORTRAN programs that
incorporate National Research Council (NRC) coefficients
and adjust the nutrient requirements for differences in
productivity and body weight (NRC 1979).
Nutrient requirements are estimated separately for
dairy versus beef cattle and for chicken layers versus
those raised primarily for meat purposes. Cattle are
divided into two populations in relation to the numbers
of breeding females classified as beef and milking
cows. The egg-laying flock is estimated from reported
data on the national average laying rate, total egg production, and meat production from chickens other than
broilers. Differences between countries in the average
laying rate and body size are incorporated in the

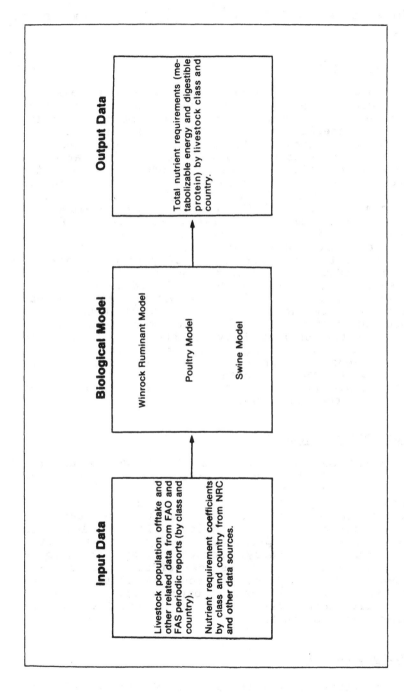

Figure 5.1 Development of base year nutrient requirement matrix

biological model for estimating nutrient requirements of poultry.

Feed Supply Matrix

Estimation of the feed supply matrix is also in terms of Mcal ME and kg DP as illustrated in Figure 5.2. This stage of assessment presented the most difficulty as data are generally not available on the feed utilization value of national pastures, crop residues, and forages. Data were also not available on the allocation of different feedstuffs to livestock production except for some former coefficient estimates of grain and protein meal use per kilogram of product (Rojko et. al. 1978).

A computer program "Feed Count" was developed in this study to estimate the nutrient contribution of concentrate feed ingredients including grains, protein meals, bran and other processing by-products (FAO 1980). The nutrient contribution of pastures, forages, and crop residues to livestock production was estimated as a residual by comparing the estimated nutrient contribution of concentrate feed use with total livestock nutrient requirements. Residual requirements were assigned to pastures, forages, and crop residues in relation to the estimated availability of these feed sources and their respective value in livestock rations. Feeds were grouped into five major classes: grains, protein meals; concentrate by-products; nonconcentrate by-products; and forages (pastures, harvested forages, and crop residues).

Livestock Offtake Matrix

Livestock offtake is an estimate of average annual production obtained from each livestock enterprise after making adjustments for changes in inventory. Both primary and secondary products are included as offtake in the case of joint products. For example, the meat offtake from dairy cattle and chicken laying flocks is treated as a secondary product. Coefficients were estimated to relate secondary product offtake to primary product offtake. All livestock offtake values are expressed in terms of metric tons as illustrated in Figure 5.3. Primary data sources for estimating national livestock offtake in the centrally planned countries were FAS and FAO reports.

Input-Output Matrix

Input-output coefficients are estimates of the energy and protein requirements from each feed group to

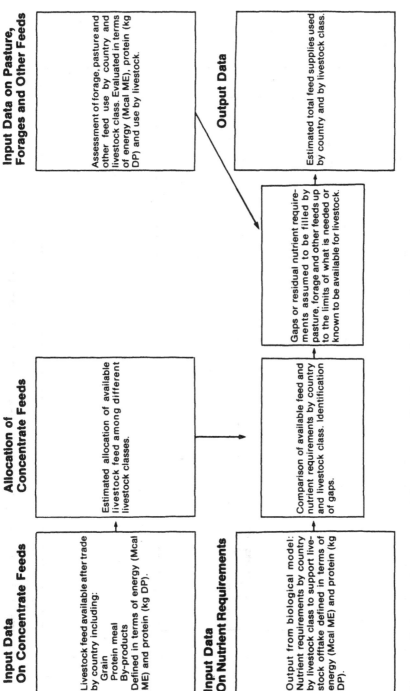

Figure 5.2 Development of base year feed supply matrix

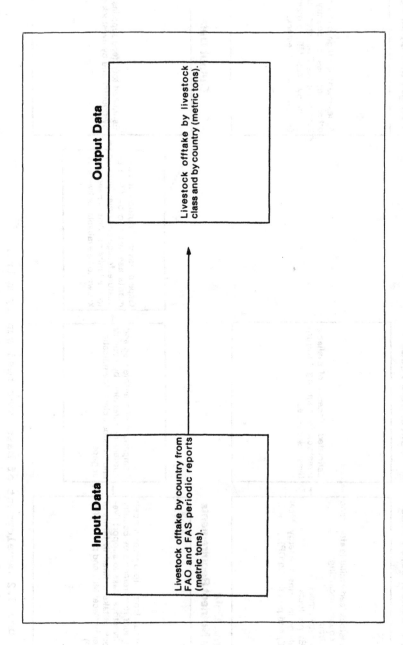

Figure 5.3 Development of base year livestock offtake matrix

produce a metric ton of livestock offtake as shown in Figure 5.4. The input-ouput matrix is based on the division, element by element, of the estimated feed supply matrix by the livestock offtake matrix. Allocations of the different feed groups to livestock classes in each country were based on general decision rules including total estimated nutrient requirements of each livestock class, information on representative rations used, and reported data on formula feed production for specialized use, for example, dairy concentrate. The allocation process was also governed by the known productivity of different livestock classes, for example, average egg-laying rate and rate of finishing for beef cattle. Roughage levels were constrained for poultry and swine rations. An iterative procedure was used in the allocation process to first exhaust the reported available supply of feedstuffs other than pastures, forages, and crop residues. As indicated under Feed Supply Matrix, the use of various roughages for feed was estimated as a residual.

In addition to feed use input-output coefficients, the input-output matrix includes joint product relationships such as milk and dairy meat. The feed use input-output coefficient indicates the feed requirement per metric ton of the primary product (milk), and the secondary product (dairy meat) is specified to be in a fixed relationship to milk offtake.

A Leontief-type impact model of the livestock feed economy in relation to domestic agricultural production, consumption and trade was used to evaluate the flow of feedstuffs through the system. The general form of this model is:

$$\begin{bmatrix} A & -I \\ I & O \end{bmatrix} \begin{bmatrix} X_1 \\ X_2 \end{bmatrix} = \begin{bmatrix} b_1 \\ b_2 \end{bmatrix} \tag{1}$$

where,

A = input-output matrix for the agricultural sector in terms of Mcal ME or kg DP per unit of output;

X_1 = domestic agricultural production vector in metric tons;

X_2 = net agricultural trade vector in metric tons;

b_1 = domestic consumption vector for agricultural products in Mcal food energy;

b_2 = policy vector for domestic agricultural production in metric tons; and

I = identity matrix.

Premultiplication of the inverse of the matrix containing A, $-I$, I, and O submatrixes in equation (1) times vector b which encompasses b_1 and b_2 leads to

132

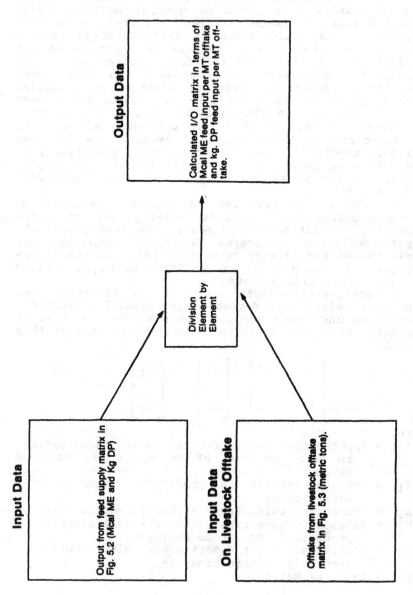

Figure 5.4 Calculation of input-output matrix for base year

Within the image:

Input Data

Output from feed supply matrix in Fig. 5.2 (Mcal ME and Kg DP)

Input Data On Livestock Offtake

Offtake from livestock offtake matrix in Fig. 5.3 (metric tons).

Division Element by Element

Output Data

Calculated I/O matrix in terms of Mcal ME feed input per MT offtake and kg. DP feed input per MT offtake.

a solution for X_1 and X_2. Values estimated for X_2 the net trade vector, are compared with actual trade data to verify operation of the model for the base period of analysis. Actual trade data may differ slightly from that estimated in equation (1) due to changes in inventory and wastage or other losses within the system. It may be noted that X_1 includes some cereal production for direct human consumption as well as animal feed in terms of metric tons; and b_1 includes cereal use for direct human consumption in Mcal ME. Losses in processing, storage, and handling also have to be accounted for in the interpretation of net trade requirements with the model.

Future trade requirements may be projected with the model on the basis of expected changes in b_2 (domestic agricultural production), and/or b_1 (domestic agricultural consumption). The most recent uses of the model for making projections at Winrock International have assumed that A, the input-output technology matrix relating feed use to livestock offtake, remains the same in making future projections (Wheeler et. al. 1981). However, the technology matrix may also be adjusted for future time periods to reflect changes in feed conversion efficiency or substitution of feed inputs for livestock production.

PROFILE OF LIVESTOCK/FEED SECTOR
IN SOCIALIST COUNTRIES

Livestock Production

Reported 1981 inventory and offtake data for livestock in centrally planned countries are shown in Table 5.1. Over 75 percent of the cattle in Eastern Europe and the Soviet Union are dual-purpose animals; whereas, cattle are raised primarily for beef production in the PRC. Estimated animal productivity is markedly lower in the PRC than the Soviet bloc due partially to the major constraints on use of concentrate feed in the PRC. The same problem applies to sheep and goats. Although nearly half of the sheep and goat population in centrally planned countries is located in the PRC, the meat offtake is only about 25 percent of that produced in other centrally planned countries. Offtake rates are also low for swine and poultry in the PRC due to the scarcity of concentrate feed. Most of the swine in the PRC are raised on private plots with low-quality rations containing a large amount of roughage and other low-quality feedstuffs. Poultry rations in the PRC contain a large amount of by-products and limited grain. In comparison to the Soviet bloc, the offtake levels of livestock produced in the PRC are estimated to be only

Table 5.1 Estimated Livestock Production in Centrally Planned Countries, 1981

Livestock	Bulgaria	Czech.	E Germany	Albania	Hungary	Poland	Yugo.	Romania	E Europe	USSR	PRC	Total CPC
Cattle												
Beef (1000 mt)												
Inventory (mil)	0.1	1.4	1.7	0.2	0.5	1.9	0.1	1.5	7.4	22.2	50.2	79.8
Dairy Inventory (mil)	1.6	3.5	4.0	0.3	1.4	9.7	5.3	4.8	30.6	92.9	14.5	138.0
Offtake Beef (1000mt)	147.0	413.0	401.0	20.0	151.0	640.0	340.0	306.0	2,418.0	6,700.0	1,678.0	10,796.0
Milk (1000mt)	1,870.0	5,980.0	8,325.0	232.0	2,653.0	15,527.0	4,402.0	4,300.0	43,289.0	90,000.0	5,395.0	138,684.0
Sheep and goats												
Inventory (mil)	10.9	1.1	2.1	1.9	3.3	3.9	7.6	16.6	47.4	147.0	180.0	374.4
Offtake Meat (1000mt)	93.0	8.0	17.0	17.0	8.0	20.0	63.0	67.0	293.0	850.0	396.0	1,539.0
Swine												
Inventory (mil)	3.8	7.9	12.9	0.1	8.2	18.7	7.9	11.5	71.0	73.4	321.6	466.0
Offtake Meat (1000mt)	373.0	833.0	1,323.0	12.0	927.0	1,433.0	762.0	915.0	6,578.0	5,200.0	15,800.0	27,578.0
Poultry												
Inventory (mil)	113.2	142.2	103.2	2.4	237.5	307.5	207.2	255.0	1,368.0	943.8	1,600.0	3,911.8
Offtake Meat (1000mt)	170.0	213.0	147.0	3.0	360.0	448.0	290.0	374.0	2,005.0	2,300.0	3,500.0	7,805.0
Eggs (mil)	2,350.0	4,900.0	5,750.0	98.0	4,500.0	8,800.0	4,550.0	6,800.0	37,748.0	69,700.0	79,920.0	187,368.0

Source: Calculated from Foreign Agriculture circulars on"Poultry and Eggs",and"Livestock and Meat",1982, Foreign Agricultural Service, USDA; and 1980 FAO Production Yearbook, Vol. 34, 1981.

about half as productive. The Chinese have made extensive use of artificial insemination and have demonstrated good management capability in other respects as evidenced by the extremely large litters of pigs. They have the potential to more than double the offtake rate of their livestock population if concentrate feed use were not so severely constrained.

Feed Use

Nutrient contributions of feeds for livestock production in centrally planned countries were estimated in terms of total metric tons fed, billion Mcal ME, and mmt DP (Table 5.2). The "Feed Count Model" derived the nutrient estimates for grain, protein meal and various by-products. Nonconcentrate by-products are distinguished from concentrate by-products in that the nonconcentrates contain relatively high fiber and moisture levels; for example, sugar beet pulp is of limited feed value to nonruminants.

Total estimated use of feed grains, protein meals, concentrate and nonconcentrate by-products in 1981 for all centrally planned countries was 224.7, 64.1, 138.0 and 42.5 mmt, respectively (Table 5.2). Protein meal use in Eastern Europe and the Soviet Union includes a high percentage of milk by-products because they are commonly used as a protein supplement in rations. Both grain and protein meal use for feed are severely restricted in the PRC compared to other centrally planned countries. Most of the available concentrate by-products in the PRC are milling offals that cannot be used for direct human consumption.

Very little information was available for centrally planned countries on the use of forages, crop residues, and grazing areas. Nutrient contributions of these roughages were estimated by comparing the difference between the total estimated nutrient requirements of livestock in each country and the nutrients furnished by grains, protein meals, and by-products.

Nutrient Requirements

Estimated annual nutrient requirements for all livestock in centrally planned countries are shown in Table 5.3. Livestock production classes include beef cattle, dairy cattle, swine, buffalo, sheep, goats, poultry, meat, eggs, horses, and camels. Mules and asses are included with horses.

Beef and dairy cattle account for about 40 percent of the estimated total feed energy requirements in centrally planned countries. Swine account for 26 percent, poultry 9 percent, sheep and goats 11 percent, and draft

Table 5.2 Estimated Annual Feed Use of Grains, Protein Meals and By-products in Centrally Planned Countries for 1980/81.

Item Nutrient	Bulgaria	Czech.	E Germany	Albania	Hungary	Poland	Yugo.	Romania	Total	USSR Total	PRC Total	CPC Total
Grains												
mmt (as fed)	5.2	8.2	8.5	0.1	8.9	18.2	9.5	13.1	71.7	121.0	32.0	224.7
DP (mmt)	0.5	0.9	0.8		0.9	1.9	0.9	1.2	7.1	12.6	3.1	22.8
ME (bil Mcal)	16.5	24.7	25.5	0.2	29.4	52.9	31.7	43.1	224.0	336.5	99.4	659.9
Protein meals												
mmt (as fed)	1.0	2.8	4.1	0.1	1.6	5.9	1.1	2.7	19.3	39.7	5.1	64.1
DP (mmt)	0.3	0.4	0.7		0.4	0.9	0.3	0.6	3.6	3.9	1.7	9.2
ME (bil Mcal)	1.4	2.6	3.9	0.1	2.1	6.0	1.5	3.2	20.8	29.6	10.2	60.6
Concentrate by-product (CBP)												
mmt (as fed)	1.5	2.7	6.0	0.2	1.1	31.0	1.8	2.8	47.1	54.2	36.7	138.0
DP (mmt)	0.2	0.2	0.3		0.1	0.3	0.2	0.2	1.5	4.1	2.3	7.9
ME (bil Mcal)	3.1	3.9	8.2	0.4	1.7	28.1	3.2	4.9	53.5	80.8	66.8	201.1
Nonconcentrate by-product (NBP)												
mmt (as fed)	0.3	1.8	1.6	0	1.6	3.8	1.4	1.5	12.0	29.3	1.2	42.5
DP (mmt)	0	0.1	0.1	0	0	0.2	0	0.3	0.7	1.0	0	1.7
ME (bil Mcal)	0.1	4.2	3.9	0	2.5	8.7	2.8	3.5	25.7	52.3	0.7	78.7
Total concentrates												
mmt (as fed)	8.0	15.5	20.2	0.4	13.2	58.9	13.8	20.1	150.0	244.2	75.0	469.3
DP (mmt)	1.0	1.6	1.9		1.4	3.3	1.4	2.3	12.9	21.6	7.1	41.6
ME (bil Mcal)	21.1	35.4	41.4	0.7	35.7	95.7	39.2	54.7	324.0	499.2	177.1	1,000.3

Table 5.3 Estimated Livestock Nutrient Requirements in Centrally Planned Countries, 1980/81

Livestock Nutrient Requirements[a]	Bulgaria	Czech.	E Germany	Albania	Hungary	Poland	Yugo.	Romania	Total	USSR Total	PRC Total	Centrally Planned Countries
Beef Cattle												
ME	522	6,596	7,916	0	2,424	8,928	0	7,036	33,422	102,269	217,269	352,960
DP	14	171	206	0	63	232	0	183	869	2,657	5,503	9,029
Dairy Cattle												
ME	7,097	19,383	21,985	2,062	7,823	53,412	22,925	21,055	155,742	404,332	63,054	623,128
DP	185	576	653	54	232	1,587	597	548	4,432	10,525	1,641	18,598
Swine												
ME	8,617	17,007	29,252	283	18,594	43,084	18,328	24,490	159,655	109,674	463,049	732,378
DP	334	650	1,135	11	722	1,672	686	957	6,167	4,257	18,125	28,549
Goats												
ME	186	61	13	664	10	242	147	347	1,670	5,454	51,558	58,682
DP	5	2	0	19	0	7	4	10	47	158	1,384	1,589
Sheep												
ME	6,262	354	2,210	925	3,088	4,071	4,386	13,559	34,855	111,470	65,309	211,634
DP	163	9	58	24	80	106	114	354	908	2,906	1,680	5,494
Poultry Meat												
ME	1,555	5,337	1,232	20	3,354	3,918	2,767	3,312	21,495	9,595	30,594	61,684
DP	80	239	63	1	172	201	142	170	1,068	492	1,569	3,129
Poultry Eggs												
ME	1,686	3,378	3,826	85	3,174	6,319	2,909	4,876	26,253	68,715	60,961	155,929
DP	70	141	160	4	132	263	121	203	1,094	2,833	2,531	6,458
Buffalo												
ME	295	0	0	11	0	0	494	1,295	2,095	2,057	170,854	175,006
DP	8	0	0	0	0	0	13	34	55	54	4,520	4,629
Horses												
ME	2,200	309	440	615	828	12,200	4,229	3,845	24,666	38,940	93,657	157,263
DP	30	4	6	8	11	163	56	51	329	521	1,289	2,139
Camels												
ME	0	0	0	0	0	0	0	0	0	1,298	3,482	4,780
DP	0	0	0	0	0	0	0	0	0	33	89	122
All Livestock												
ME	28,420	52,425	66,874	4,665	39,295	132,174	56,185	79,815	459,853	853,804	1,219,787	2,533,444
DP	889	1,792	2,281	121	1,412	4,231	1,733	2,510	14,969	24,436	38,331	77,736

[a]ME in million Mcal DP in 1000 mt. Horses include horses, asses, and mules.

animals 14 percent. Total nutrient requirements in 1981
were estimated to be 2,533 billion Mcal ME and 78 mmt DP
for all centrally planned countries. Dairy cattle
account for about two-thirds of all cattle requirements,
and eggs about the same percentage of all poultry requirements.

Distribution of the feed-energy requirements for
livestock was 18 percent in Eastern Europe, 35 percent
in the Soviet Union and 47 percent in the PRC (Table
5.3). Poland is the principal livestock-producing
country in Eastern Europe, accounting for 26 percent of
that region's feed-energy requirements.

RESULTS OF THE STUDY

Projected Grain Use

The estimated grain use in the Soviet Union, Eastern
Europe, and China derived from the input-output model is
shown in Table 5.4. Feed use in 1980-1982 averaged 224
million tons, about 54 percent of production. Feed use
of grain is expected to increase to 351.8 million tons
or between 53 percent and 67 percent of expected production in the year 2000.

Soviet Union

Dairy, pork, and poultry are the major grain users
in the Soviet Union. About 169.2 mmt of grain and an
increase in grain fed of 48.8 percent will be required
to produce FAO projections of 21.4 mmt of meat as well
as milk and egg requirements in 2000. Domestic
production of grain forecast for 1990 and 2000 is based
on the recent historical trend and projections reported
by R.O. Wheeler (Wheeler 1983). The trend estimate for
the Soviet Union is 0.3 percent and the alternative
projection is 2.3 percent per year. The implied grain
import requirement of meeting these projections,
assuming poor domestic grain crops reflecting the recent
trend, is about 94.9 mmt. On the other hand, if grain
production increases at 2.3 percent per year, this
reduces grain import needs in 2000 to about 38.3 mmt
compared with 39.3 mmt in the 1980-82 period. These
estimates assume noncereal animal-feed input-output
coefficients and meat imports will remain the same as
current levels. In the case of the Soviet Union,
long-term grain imports can continue to be purchased
through gold production and energy exports.

On May 24, 1982, the Soviet Union approved a "Food
Program" to provide a coordinated marketing effort from
the farm to consumer. This effort should improve
management and price incentives in the food system.

TABLE 5.4
Projected Wheat and Coarse Grain Production, Utilization, and Trade Requirements
for Grain 1980-1982 Average and 2000

	1980-1982	1990		2000	
		thousands of metric tons			
Soviet Union					
Production	167,333	171,906	205,335	177,133	257,762
Growth rate[a]		(0.3%)	(2.3%)	(0.3%)	(2.3%)
Utilization	206,667	244,792	254,781	271,996	296,089
Food	43,000	46,870	46,870	49,880	49,880
Feed[b]	113,667	146,556	146,556	169,188	169,188
Other	50,000	51,366	61,355	52,928	77,021
Trade balance	39,334	66,886	49,446	94,863	38,327
China					
Production	146,900	180,261	191,671	226,287	257,590
Growth rate[a]		(2.3%)	(3.0%)	(2.3%)	(3%)
Utilization	165,979	204,934	204,934	249,521	249,521
Feed[b]	31,900	52,084	52,084	72,269	72,269
Other	134,079	152,850	152,850	177,252	177,252
Trade balance	19,079	24,673	13,263	23,234	8,069
Eastern Europe					
Production	98,607	108,810	121,000	121,389	151,389
Growth rate[a]		(1.1%)	(2.3%)	(1.1%)	(2.3%)
Utilization	106,933	132,208	132,208	153,799	153,799
Feed[b]	68,400	90,978	90,978	110,257	110,257
Other	38,533	41,230	41,230	43,542	43,542
Trade balance	7,900	23,398	11,208	31,410	1,904

a The annual growth rate in grain production is based on historical and projected growth
rates reported in R.O. Wheeler, World Agriculture: Review and Prospects into the 1990s.
Winrock International, Morrilton, Arkansas, 1983.
b Livestock output projections to 1990 and 2000 are based on preliminary FAO projections
for their AD 2000 report.

D. Gale Johnson has contended that the major agricultural problem in the Soviet Union is as much a problem of policy as climate (Johnson 1982).

The Soviet five-year plan for 1986-1990 appears optimistic relative to the former projections (Table 5.5). The planned production of all grains of 252 mmt would require an annual growth rate of about 5 percent per year. Even if the Soviet Union could conceivably increase grain production at this rate, they would still require about 22 mmt of grain imports to attain their meat production goals. The basic projection data and their plans show consistent results. Planned meat production is targeted to increase 17.4 percent in the 1986-1990 five-year plan, whereas planned grain production is only 5 percent. The Soviet Union will continue to import sizable quantities of grain, even to meet modest increases in meat production.

Eastern Europe

Serious financial problems in Eastern Europe caused by large trade deficits cloud any current analysis of grain imports. As debt-servicing adjustments occur, the economic growth of the region will probably decline somewhat. Much caution should be used in reviewing projections in this region, since internal policy in the short run could change import needs substantially.

Hogs, dairy, and poultry are the major livestock groups fed in Eastern Europe. In 1980-1982, 68.4 mmt were fed constituting about 69 percent of wheat and coarse-grain production. Grain feeding is expected to increase to 110.3 mmt in 2000 taking 73 percent and 91 percent of the alternative production projections. If meat production increases from 10.1 to 17.8 mmt by the year 2000 and grain production grows over the same time period by 1.1 percent annually, grain import requirements will be 31.4 mmt. With a higher production growth rate at 2.3 percent per year, import needs will be only 1.9 mmt.

Eastern Europe is presently going through a volatile, political, and economic period. The USDA has determined that the agricultural import market has declined in recent years and "prospects for recovery are dim through the eighties." "Rising debt-service ratios and increasing difficulty in obtaining new credit are major causes of the turnaround." (Cook et. al. 1984).

China

Poultry and hogs use most of the grain fed in China. FAO projections indicate meat production at 32.4 mmt in the year 2000. The grain requirement to pro-

TABLE 5.5
Soviet Union Meat and Grain Production and Plans (mmt) 1961-1990.

	Actual				Plans	
	1961-65	1966-70	1971-75	1976-80	1981-85	1986-90
Meat production	9.3	11.6	14.0	14.8	17.25	20.25
Grain production	130.3	167.6	181.6	205	240.5	252.2
Grain for feed	40.2	69.3	97.4	120.8	138	159
Percentage increase in meat production		24.7	20.7	5.7	16.6	17.4
Percentage increase in grain production		28.7	8.4	12.9	17.3	5.0
Percentage increase in grain fed		72.4	40.5	24.0	14.2	15.2

TABLE 5.6
People's Republic of China Meat and Grain Production and Plans (mmt) 1985-1990.

	Actual		Plans	
	1977-1979	1980	1985	1990
Meat production	9.2	12.10	14.0	18.5
Grain production	306.54	318.20	375	425
Percentage change in meat production		31.5	15.7	32.1
Percentage change in grain production		3.8	17.9	13.3

duce that amount of meat as well as projected milk and
eggs is about 72.3 mmt compared with only 31.9 mmt in
1980-1982. China was feeding about 22 percent of its
wheat and coarse-grain production in 1980-1982. This
would increase to about 28 percent by 2000. If grain
production increases at an annual rate of 2.3 percent
per year, projected grain import needs in 2000 are about
23.2 mmt. However, if production is increased at 3 percent per year, grain imports in 2000 could be reduced to
about 8.1 mmt.

The production plans for the Peoples' Republic of
China are shown in Table 5.6.

The planned total grain production in 1985 was much
more than many observers thought was possible. The USDA
projected 375 mmt total grain production (including
rice), which gave them a net (excluding offals) production of about 280 mmt. If human consumption is on target, food use of grain will equal total producton of
about 280 mmt. (Kilpatrick 1982).

Therefore, the grain import requirement to meet projected meat production in 1990 would be approximately 28
million tons. Targeted meat production is to increase
by 32 percent and grain production 13 percent. However,
"China's agricultural import policy remains conservative
and from the limited indications available shows no sign
of evolving in the direction of accepting rising levels
of imports and growing participation in international
specialization as a desirable goal." (Surls 1982).

Total Grain Fed

Estimated grain fed in the centrally planned countries derived from the input-output model compares
closely with USDA feed estimates even during the period
of the U.S. grain embargo, when normal feeding conditions were difficult to maintain (Table 5.7).

Using FAO projections for meat, dairy, and egg production to the year 2000 and the model input-output
coefficients, projected grain feeding requirements are
169.2 mmt in the Soviet Union, 110.3 mmt in Eastern
Europe, and 72.3 mmt in China. FAO's own estimates of
feed use are lower in the Soviet Union and Eastern
Europe but they do not account for grain use by other
livestock such as horses and mules. The FAO feed estimate in China is higher than the estimate from the model
since the FAO estimate is based on an assumed change in
the feed diet particularly in the swine industry.

Analysis of Oilseed Feed Use

Both the Soviet Union and Eastern Europe have
attempted to increase the use of protein in livestock

TABLE 5.7
Model Feeding Results for the Centrally Planned Countries (mmt) 1979 to 1983 and Projected to 2000.

	Soviet Union		Eastern Europe		China
	USDA estimates[a]	Model estimates	USDA estimates	Model estimates[a]	Model estimates
1979	69.4	69.9	123	121.5	28.9
1980	71.7	71.7	119	119	30.3
1981	65.8	68.4	116	115.1	31.9
1982	69.7	69.0	117	115.4	33.4
1983	66.3	66.2	123	118.5	34.9
Model 2000		110.3		169.2	72.3
FAO 2000[b]		103.3		159.2	86.0

[a]USDA Foreign Agriculture circular: "Grains," Washington, D.C.
[b]FAO 2000 Projections, unpublished working papers.

feed rations over the past few years in order to improve feed conversion efficiency. In former years, grain was commonly used as the only concentrate feed supplement. Estimated 1981 feed use of oilseed meals (the major protein supplement), is shown in Tables 5.8, 5.9, and 5.10 for the three centrally planned regions.

Use of oilseed meal in livestock rations is projected for 1990 and 2000 on the basis of estimated input-output coefficients for 1981 and alternative assumptions regarding the future growth of domestic production and other competing uses for oilseeds. In the case of the Soviet Union (Table 5.8), the higher production growth rate is 2.9 percent per year as estimated in World Agriculture while the lower growth rate is only 0.5 percent per year, reflecting the recent historic trend. Livestock producton in 1990 and 2000 is based on the Food and Agricultural Organization (FAO) projections. Increases in oilseed meal requirements are approximately equal to increases in livestock production and are projected to rise from 6.2 mmt in 1981 to 9.4 mmt in 2000. The Soviet Union presently uses a substantial amount of milk and milk by-products for protein supplement, and the availability of these other protein feeds is assumed to remain about the same as in 1981 at 2.6 mmt soybean meal equivalent.

Results for the Soviet Union indicate a significant reduction in import requirements for oilseeds in the period from 1981 to 2000 if the future domestic production of oilseeds expands at the rate projected in World Agriculture, i.e., at a rate of 2.9 percent per year. On the other hand, if production continues at the recent trend level of only 0.5 percent, import requirements are projected to increase from 2.8 mmt oilseed meal equivalent in 1981 to 5.2 mmt in 2000. Increased use of oilseeds as a component in livestock rations would raise the import requirements further.

Projections of oilseed requirements in Eastern Europe shown in Table 5.9 are based on two growth rates for domestic production -- 2.0 percent and 3.5 percent per year. The higher rate is forecast in World Agriculture (USDA, FAS 1983). The lower rate used here compares with a recent trend rate of 3.5 percent; however, the potential for further expansion of oilseed production is limited according to sources in USDA Foreign Agricultural Service. Irrespective of the production growth rate assumed, the trade requirements for oilseeds in Eastern Europe are projected to increase after 1981 as indicated in Table 5.9. Improvements in ration formulation utilizing more protein meal components, as occurred in Western Europe, would increase the requirements above that forecast here with estimated 1981 input-output coefficients.

Oilseed projections for China shown in Table 5.10 are based on a 3 percent growth rate for production and

TABLE 5.8
Projected Production, Utilization and Trade Requirements for Oilseeds in the Soviet Union, 1981-2000.

Item	1981	1990		2000	
		------------(1000 mt)------------			
Domestic production:					
Oilseeds	10,246	10,716	15,744	11,264	21,255
Growth rate[a]		(0.5%)	(2.9%)	(0.5%)	(2.9%)
Oilseed meal[b]	3,910	4,032	6,645	4,218	9,410
Other protein feeds[c]	2,603	2,603	2,603	2,603	2,603
Domestic utilization:					
Feed requirements:					
Oilseed meal	6,187	7,813	7,813	9,440	9,440
Other protein feed	2,603	2,603	2,603	2,603	2,603
Other requirements:[d]					
Oilseeds	2,714	2,958	2,958	3,148	3,148
Trade requirements:					
Oilseed meal	2,277	3,781	1,168	5,222	30
Oilseed equivalent[e]	2,892	4,802	1,483	6,632	38

[a]The lower growth rate is based on an 0.5% annual increase according to past trends in production. The higher growth rate is based on a 2.9% annual increase in production as forecast in Winrock International, World Agriculture.

[b]Oilseed meal produced from domestic oilseed supply.

[c]Other protein feeds are composed primarily of milk fed.

[d]Other oilseed requirements include waste, seed, and human consumption.

[e]Oilseed meal requirement in whole soybeans.

TABLE 5.9
Projected Production, Utilization, and Trade Requirements for Oilseeds in Eastern Europe, 1981-2000.

Item	1981	1990		2000	
		------(1000 mt)------			
Domestic production:					
Oilseeds	3,901	4,662	5,697	5,683	8,070
Growth rate[a]		(2%)	(3.6%)	(2%)	(3.6%)
Oilseed meal[b]	1,880	2,268	2,812	2,795	4,049
Other protein feeds[c]	2,419	2,149	2,149	2,149	2,149
Domestic utilization:					
Feed requirements:					
Oilseed meal	7,193	10,030		12,868	
Other protein feed[d]	2,149	2,149		2,149	
Other requirements:					
Oilseeds	324	347		366	
Trade requirements:					
Oilseed meal	5,313	7,762	7,218	10,073	8,819
Oilseed meal equivalent[e]	6,748	9,858	9,167	12,793	11,200

a The lower growth rate is based on a production growth rate of 2%/year compared with a 3.5% past trend rate. The higher growth is based on a 3.6% future growth rate as fore-cast in Winrock International, World Agriculture.
b Oilseed meal produced from domestic oilseed supply.
c Other protein feeds are composed primarily of milk fed.
d Other oilseed requirements include waste, seed, and human consumption.
e Oilseed meal requirement in whole soybeans.

TABLE 5.10
Projected Production, Utilization, and Trade Requirements for Oilseeds in China, 1981-2000.

Item	1981	1990		2000	
		Scenario 1[a]	Scenario 2[b]	Scenario 1[a]	Scenario 2[b]
		(1000 mt soybean equivalent)			
Domestic production:					
Oilseeds	19,561	25,523	25,523	34,300	34,300
Oilseed meal[c]	3,847	5,520		8,063	
Other protein feeds[d]	80	80	80	80	80
Domestic utilization:					
Feed requirements:					
Oilseed meal	4,400	7,101	7,101	9,802	9,802
Other protein feeds	80	80	80	80	80
Other requirements:					
Oilseeds for food	4,845	5,524	5,524	6,395	6,395
Oilseeds for nonfood	8,576	11,189	3,816	15,037	5,128
Trade requirements:					
Oilseed meal	553	1,581	(4,224)	1,739	(6,063)
Oilseed equivalent	702	2,008	(5,365)	2,209	(7,700)

a Scenario 1 assumes that nonfood use of oilseeds for largely fertilizer use will increase 3% annually in relation to production.
b Scenario 2 assumes that use of oilseeds for fertilizer will be discontinued over time.
c Oilseed meal produced from domestic oilseed supply.
d Other protein feeds are composed primarily of milk fed.
e Oilseed meal requirement in whole soybeans.

two alternative scenarios for oilseed use. Due to current fertilizer shortages in the PRC, much of the available oilseed and meal supply is used for fertilizer. About 8.6 mmt of oilseeds in 1981, identified in Table 5.10 as oilseeds nonfood, included a substantial amount of fertilizer use. Scenario 2 assumes that oilseed use for fertilizer will be discontinued by 1990 and remaining nonfood uses for seed and waste will continue in proportion to increases in oilseed production. Scenario 1 assumes that all nonfood uses of oilseeds will continue to increase in proportion to production. Under Scenario 2, China is projected to develop a major surplus of oilseeds by 1990, increasing to 7.7 mmt by 2000. Under Scenario 1, China would incur a 2 mmt deficit in 1990 and 2000.

Use of By-Products and Roughage

By-products used for feed are separated into two categories. Concentrate by-products, including various cereal brans and grain substitutes such as dried cassava, may be used to formulate concentrate feeds for all livestock. Nonconcentrate by-products; including fresh cassava, waste bananas, and citrus pulp, have either a high water content and (or) a high fiber content thus restricting their use in feeding, in particular, for nonruminant species. The major source of data for determining feed use of different by-products was the 1975-1977 FAO food balance sheets. Production of concentrate by-products is closely associated with processing of food grains and is expected to level off as human diets are improved.

Feed use of pasture, hay, silage, and other roughages is determined as a residual by comparing the nutrient contribution of all other feeds with a known supply against the total nutrient requirement of the national livestock population.

The estimated nutrient contribution in billion Mcal of various feedstuffs in 1981 was as follows: 1) Soviet Union; grain (336.5), protein meal (29.6), concentrate by-products (80.8), nonconcentrate by-products (52.3), and roughages (367.4); 2) Eastern Europe: grain (224.0), protein meal (20.8), concentrate by-products (53.5), nonconcentrate by-products (25.6), and roughages (118.5); and 3) China: grain (99.4), protein meal (10.2), concentrate by-products (66.8), nonconcentrate by-products (0.7), and roughages (967.5). Roughages comprise a much higher proportion of feed supply in China than in other centrally planned countries. All centrally planned countries could expand use of roughages with improved pasture management, increased use of production inputs, and increased utilization of crop residues.

CONCLUSIONS AND IMPLICATIONS

An input-output energy flow model is used to quantitatively evaluate interactions in the livestock and grain system in the centrally planned countries. Livestock production is expected to increase substantially in these countries and as a result continue to have a large impact on the world market for grains and oilseeds. This model is a very useful model even though it is a static model and assumes fixed input-output coefficients between livestock output and feed inputs. Given present information on the centrally planned countries, it is difficult to take into account changes in price relationships and changes in technology. However, as these data become available, one could easily interface this model with optimization models.

This model is useful in estimating short-term use in the centrally planned countries because data are limited and the output goals of these countries are known in advance. Exogenously any changes in government plans can be incorporated into the model and the resultant pull on oilseeds, grains, and other feedstuffs can be analyzed. Since these countries will have a substantial impact on the world grain and oilseed market to the year 2000, this modeling effort is a first attempt at specifying feeding relationships. As more information becomes available on the grain-livestock feed relationships, especially for forages and crop residues, the model can be enlarged to handle this important segment of the industry where additional research is necessary.

The grain-oilseed-livestock relationships reported in this chapter assume normal political relationships, relatively stable grain/livestock prices, consistent internal policies, and normal climatic conditions. Changes in these factors could have a significant impact on the results reported.

The major expansion in grain and oilseed trade in recent years is attributed to increased emphasis on livestock production and the shift from traditional to modern feeding practices. This trend in livestock production and improved ration formulation has important implications for major grain and oilseed exporters such as the United States. By being able to forecast grain and oilseed use more effectively, this enables the USDA to implement more appropriate farm policy and planned resource utilization. The objective is to maintain a high degree of responsiveness to changing world market conditions and thus avoid periodic disruptions such as occurred in the grain market in 1967, 1972-1974, and again in 1980.

Since the centrally planned countries represent an important growing market for oilseeds and grain, the United States should develop trade policies to facilitate increased trade. This might include increased

150

technical assistance in livestock production, ration formulation, promotion of feed trade, and increased credit availability. In addition, special attention must be given to providing assistance in developing their own export markets for products in which they have a comparative advantage to improve their foreign exchange position. Ultimately this assistance would benefit the United States since the United States has a relative advantage in the production of grains and oil-seeds.

NOTES

1. This estimate was reported by the Food and Agricultural Organization (FAO) Committee on Commodity Problems: Intergovernmental Group on Grains at their 20th Session in Rome 1980. No official estimates are available from the USDA.

REFERENCES

Cook, E., R. Cummings, and T.A. Vankai. Eastern Europe: Agricultural Production and Trade Prospects through 1990. Washington, D.C.: ERS, USDA, Foreign Agricultural Economics Report No. 195 (February 1984).

Cramer, G. L. and C. W. Jensen. Agricultural Economics and Agribusiness. Chapter 13. New York: John Wiley and Sons, Inc. (1984).

Cramer, G. L. and W. G. Heid, Jr. (ed.) Grain Marketing Economics. New York: John Wiley and Sons, Inc. (1983).

Fitzhugh, H.A., H.J. Hodgson, O.J. Scoville, T.D. Nguyen and T.C. Byerly. The Role of Ruminants in Support of Man. Morrilton, Arkansas: Winrock International (1978).

Food and Agricultural Organization. Food Balance Sheets, 1975-77 Average. Rome: United Nations (1980).

Joint Economic Committee, Congress of the United States, 97th Congress, Second Session. China Under Four Modernizations, Selected Papers, Part 1: Washington, D.C.: U.S. Government Printing Office (December 1982).

_____. Congress of the United States, 97th Congress, Second Session. Soviet Economy in the 1980s: Problems and Prospects, Selected Papers, Part 2: Washington, D.C.: U.S. Government Printing Office (December 1982).

Joint Economic Committee, Congress of the United States, 97th Congress, First Session. East European Economic Assessment. A Compendium of Papers, Part 2. Washington, D.C.: U.S. Government Printing Office, (July 1981).

_____. Congress of the United States, 97th Congress, Second Session. China Under Four Modernizations, Selected Papers, Part 2. Washington, D.C.: U.S. Government Printing Office (December 1982).

_____. Congress of the United States, 97th Congress, Second Session. Soviet Economy in the 1980s: Problems and Prospects, Selected Papers, Part 1. Washington, D.C.: U.S. Government Printing Office (December 1982).

Kilpatrick, J. A. "China: the Drive for Dietary Improvement." In: China Under Four Modernizations. Part 1. Joint Economic Committee, Congress of the United States, 97th Congress, 2nd Session. Washington, D.C.: U.S. Government Printing Office. (August 13, 1982).

Nguyen T.D. and H.A. Fitzhugh. Winrock Model for Simulating Ruminant Production Systems. Morrilton, Arkansas: Winrock International.

Regier, D. Livestock and Derived Feed Demand in the World GOL Model. Washington, D.C.: ESCS, USDA. (September 1978).

Rojko, A., D. Regier, P. O'Brien, A. Coffing and L. Bailey. Alternative Futures for World Food in 1985. Washington, D.C.: USDA. (April 1978).

National Research Council, Nutrient Requirements of Domestic Animals, No. 5. Nutrient Requirements of Sheep. Washington, D.C.: National Academy of Sciences-National Research Council. (1975).

_____. Nutrient Requirements of Domestic Animals, No. 1. Nutrient Requirements of Poultry. Washington, D.C.: National Academy of Sciences-National Research Council. (1977).

_____. Nutrient Requirements of Domestic Animals, No. 3. Nutrient Requirements of Dairy Cattle. Washington, D.C.: National Academy of Sciences-National Research Council. (1978).

_____. Nutrient Requirements of Domestic Animals, No. 2. Nutrient Requirements of Swine. Washington, D.C.: National Academy of Sciences-National Research Council. (1979).

Surls, F.M. "China Grain Imports." In: China Under Four Modernizations, Part 2. Joint Economic Committee, Congress of the United States, 97th Congress, 2nd Session. Washington, D.C.: U.S. Government Printing Office. (August 13, 1982).

Treml, V.G. "Subsidies in Soviet Agriculture: Record and Prospects." In: Soviet Economy in the 1980s: Problems and Prospects; Part 2. Joint Economic Committee, Congress of the United States, 97th Congress. Washington, D.C.: U.S. Government Printing Office. (December 1983): 171-185.

USDA, F.A.S. "World Oilseed Situation and U.S. Export Opportunities." Washington, D.C.: Foreign Agriculture Circular. FOP 6-83. (June 1983).

152

USDA, Foreign Agricultural Service. "Grains", Washington, D.C.: Foreign Agriculture Circular (FG-31-83): 11-12.

Wheeler, R. O. World Agriculture: Review of Prospects into the 1990s. Morrilton, Arkansas: Winrock International (December 1983).

Wheeler, R. O., G. L. Cramer, K. B. Young and E. Ospina. The World Livestock Product, Feedstuff, and Food Grain System. Morrilton, Arkansas: Winrock International (1981).

6

China: An Enigma in the World Grain Trade

C. Peter Timmer and James R. Jones

INTRODUCTION

Until recently Chinas' efforts to feed its people after the establishment of the PRC regime in 1949 have been perceived as largely successful by the standards of both its own history and by the comparative experience of its large Asian neighbors, especially India, Indonesia, and the Philippines. More recent evidence challenges this view as new data and analysis reveals that the egalitarian policies of Mao led in many instances of reduced per capita food consumption, increases in the mortality rate and widespread cases of hunger, especially in the aftermath of the excesses of the "Great Leap Forward". Nevertheless, the Maoist experiment with egalitarian food distribution mechanisms and the post Mao reforms which have ushered in the responsibility system that stresses material incentives, have been watched closely by the rest of the world for several reasons. China's success could have broad lessons for other countries' development strategies. More to the point of this book, as an actor in the world food system, China will be watched for its potential impact on agricultural trade and prices. Since China feeds nearly a fourth of mankind, trends and shifts in its demand for food relative to supply can have significant spillover effects on the rest of the world food system. Whether China is like other centrally planned economies or whether Deng Xiaoping's sweeping economic reforms will erase most similarities to centrally planned socialism, its size will require monitoring the effect of its modernization program on the world food economy.

Other developing countries have managed the transition to modest affluence by increasingly resorting to agricultural specialization and international trade. China is currently in the position that Japan, South Korea, and Taiwan were in some years ago as they became progressively more affluent and imported substantial quantities of wheat and feed grains to satisfy new, income-led food demand patterns. Japan has become one of the world's largest importers of grain, despite its much publicized success in reaching self-sufficiency in rice. Taiwan and South Korea are major importers now also. Were China to follow a similar development path with increasing reliance on international grain markets for wheat and feed grains, the ramifications for the world food system would be major and complicated. China could affect international markets on a scale as large or larger than the Soviet Union and Eastern Europe did when they entered the grain trade as major importers to upgrade the diets of their populaces in the 1970s. Simultaneously, the necessary adjustments in international grain markets would feed back to influence the costs and potential success of China's development strategy.

After a brief review of Chinese food policy, we examine the current dilemma over future food policy directions. Concerns for production stagnation, the role of incentives, available policy instruments for maintaining high farm prices and low food prices, and the guarantee of minimum food consumption floors are not unique to China. But China's sheer size presents a new element to the possible role of international markets as an ingredient in a successful food policy framework.

Consequently, part of the discussion is devoted to understanding the mechanisms and paths that connect the Chinese food economy to international grain markets. In the 1970s these markets became progressively less stable and increasingly dependent on North America for additional supplies. Lessons gained during this period brought out the interdependence between domestic food policies of emerging middle income and centrally planned economies on the one hand and international markets on the other. What is in store for the remainder of the century could depend heavily on China. Among the forces that will provide the dynamics of this interconnected system will be; 1) growth in China's population; 2) growth in incomes, if and as China reaches its aspiration to become a middle income country; and 3) China's need and ability to enter international markets to supplement its food supplies.

It is important to consider China's potential impact in the food grain sector. China has traditionally used the international grain market to import cheap wheat and export expensive rice. Whether this calorie arbitrage reflects a conscious effort to maximize foreign exchange earnings or minimize logistical problems or both is debatable. The significant point is that reliance on world markets for an increasing proportion of wheat consumption needs seems to reflect an increased willingness to judge the domestic resource cost of producing additional grain against some international price standard. There is already evidence that one possible outcome of China's modernization and liberalization policies is to shift emphasis from food grain production to specialty and cash crops.

Another important consideration is that income growth could create demand for meat and livestock products that can be met only by increased feeding of grain, especially corn, to livestock.

China is not a "small country" in this exercise, and so it cannot assume that the international market conditions are unchanged by its own domestic policy decisions. The chapter examines this simultaneous relationship between Chinese development policy as it relies on income incentives to increase productivity, resulting changes in demand on world grain markets because of the need to produce more food including grain-fed livestock products, and the financial and foreign exchange costs of maintaining the income incentives in real terms. Hence the efficacy of much of the Chinese development strategy, to achieve modernization by the end of the century, may hinge on the responsiveness of the world grain market to possible Chinese demand for wheat and feed grains.

CHINESE FOOD POLICY

Only a brief review of the basic elements of Chinese food policy is presented here to provide an appropriate context for the current policy dilemmas that have emerged since the death of Chairman Mao. The starting point is the revolutionary restructuring of ownership and decision making patterns in the countryside that culminated in the communal organization of most peasants in 1958. Much experimentation had gone on with this organizational form, as the appropriate balance was sought between incentives to individual households and workers relative to a broadened base of shared output to produce more egalitarian income distribution. No satisfactory answer has been found, and the current debate continues to focus on this issue. Just when the basic communal form of peasant organization and ownership seemed firmly rooted, the inauguration and implementation of the responsibility contract system dramatically altered that system.

Production of agricultural output from China's communes only slightly outpaced population growth during the first thirty years of Communist rule, and significant regional differences existed in even this modest performance. With the emphasis from 1967 to 1977 on increased regional self-sufficiency in basic grain production, regions poorly suited to growing wheat, rice, or corn fared relatively poorly. Even Sichuan Province, traditionally a grain surplus area, suffered

production declines in the face of economically inefficient grain production patterns mandated by State
planners.

Apart from production declines and stagnation during
years of significant economic disruption -- the Great
Leap years of 1959-61, the Cultural Revolution years of
1967-69, and Gang of Four years of 1975-77 -- aggregate
grain production responded to investments in water
control, both irrigation and drainage, to improved seed
technology, and to the improved availability of chemical
fertilizers. With much agricultural decision making
centralized at the production team or brigade level, and
planning coordinated at the commune level, the diffusion
of new agricultural technology made impressive strides.
By the end of the 1970s Chinese agriculture appeared to
have achieved a relatively high productivity base compared to other developing countries. For example, comparison of Chinese and Indian agricultural performance
in the late 1970s showed China producing a quarter more
grain per capita for half again as many people from
three-quarters of the arable land. On the other hand,
it has been noted that China's grain yield lags behind
that of Japan, Taiwan and North and South Korea (Hsu
1982). Moreover the costs of increasing grain production became consistently greater with the result that
per capita incomes of rural producers declined and in
some cases the dislocation caused by the Maoist policies
led to extreme hardship and even hunger.

Much of the world's interest in Chinese food policy
during the first thirty years of Communist administration stemmed from the perceived ability of China to
guarantee minimum food consumption levels to all strata
of society, especially to both urban and rural poor.
During the 1970s both naive and informed travelers to
Chinese cities and successful rural areas uniformly
reported an absence of the severe hunger and malnutrition that was all too apparent in other Asian cities and
countrysides. As China has opened to the outside and
more information has become available, it has become
apparent however that China had not solved its food
problem. The Chinese themselves have indicated that
perhaps one hundred million individuals, primarily in
the poorer, outlying provinces and regions, were not
getting enough food to eat. Indirect evidence of
increased mortality rates during the Great Leap years
suggests that widespread famine conditions must have
existed (Lardy, 1984). Not everyone in China was adequately fed, and relatively few had a diet they themselves would judge satisfactory in quality and variety.

With all the provisos and exceptions taken, however,
the fact remains that China's command economy had provided a very large proportion of its very poor population with a diet adequate for normal physical activity
and growth. Infant mortality in China, always a sensi-

tive indicator of the nutritional status of the poor, became low relative to other poor countries (but not the lowest, as Sri Lanka and South Korea indicate). Life expectancy has risen sharply since 1949. For its average income and calorie availability levels during the Mao era, the Chinese may have achieved more even food distribution and resulting nutritional status than any other large poor country.

The food distribution mechanisms by which this was accomplished involved carefully designed and controlled food rations at very low prices in urban centers and food grain distributions based partly on need and partly on work points earned through labor in rural areas. Both urban and rural distribution mechanisms were designed specifically to prevent relatively more wealthy citizens in China -- China's income distribution was not noticeably more even than Taiwan's or Sri Lanka's -- from bidding basic foodgrains away from poor consumers. Foodgrain was cheap to consumers but not in unlimited quantities. These egalitarian policies were significant in that otherwise more would have been consumed by the wealthy directly, and more grain would have been fed to livestock and consumed indirectly as meat. The poorest parts of China's population would have had to supply this added grain from their own consumption if market prices allocated grain to the highest bidder. The urban and rural rationing systems were designed to prevent this skewing of foodgrain intake, and on this count they were remarkably successful.

Implementing such a system during the Mao years was not costless. In addition to the bureaucratic resources needed for day to day operations, the maintenance of low grain prices to consumers requires either low prices to producers, with the attendant depressing effects on incentives to produce, or substantial subsidies to pay the marketing costs between producer and consumer. China has used elements of both strategies, with urban grain prices actually approximately the same as, or slightly below, prices paid to communes for above-quota grain sales before 1979 and well below procurement prices today.[1]

Hence the marketing subsidy was so large as to render the marketing margin zero. At the same time, Chinese farmers received relatively low prices for their grain sales within state quotas compared either to prices they paid for agricultural inputs, such as fertilizer and machinery, or to prices received by other Asian farmers for similar commodities. The fact that the terms of trade for Chinese agriculture improved significantly during the Mao years did not alter the fact that they still discriminated significantly against agriculture. Raising of prices for above-quota sales and opening of rural markets to grain sales during the post Mao period which are discussed below appear to be steps toward removing this bias. However the shift towards price incentives and decentralized decision making opens a new dilemma. Higher food prices and greater incentives to produce industrial and cash crops have inevitably skewed income distribution in China's rural areas, while lowering the urban-rural disparity. This trade-off could be exacerbated if drains on the state treasury from higher food prices coupled with subsidized prices to consumers require policy adjustments. China is a low income country. How it feeds itself presents a dilemma that, while also shared by the European centrally planned economies, is critical to its very existence.

CHINA'S FOOD POLICY DILEMMA
AND MODERNIZATION DRIVE

A difficult dilemma shared by many countries, not just China, occurs over basic directions for their domestic food policies. There is a constant struggle to keep agricultural productivity in pace with domestic food consumption requirements. Higher food prices to provide added incentives to farmers to invest in short-run and long-run techniques for greater food production can cause food consumers great distress, especially those urban and rural landless food consumers who must purchase all their food. Poorer food consumers suffer from higher food prices the most. A high food price policy is widely understood to lead to sharply skewed food intake by income class, especially when higher-income consumers are able to purchase meat from grain-fed livestock.

China is likely to face this basic food policy dilemma in the next decade. Production has spurted upward in the 1980s apparently in response to a set of new incentives to peasants, which have included higher prices and significantly greater freedom to allocate their resources, including land, to those crops and techniques that would maximize net revenue, and because of increased availability of fertilizer and other inputs. Because some of these incentive schemes are a one shot event and prices cannot be increased at the rural level relative to urban retail prices indefinitely, debate continues about the pressure that resource constraints and limited new agricultural technology available during the coming decade will place on potential growth in Chinese agricultural output.

One school of foreign scholars has consistently argued that China's agricultural growth has been retarded by a lack of incentives to produce and much too heavy an involvement by central and provincial planners in allocating specific crops to specific areas. As part of a drive to increase local and regional self-reliance, too much acreage was devoted to basic grains and not enough to more productive cash and industrial crops. Permitting greater attention to comparative advantage through an active state role in guaranteeing grain availability to grain-deficit rural areas producing non-grain crops can be a source of significant and sustained growth in agricultural output. Furthermore such a strategy would be much more successful if the traditional bias in the terms of trade against agriculture, and especially against cash and industrial crops, were reversed through an improvement in output prices and lower costs for modern inputs, especially fertilizer and machinery. The resulting improvement in incentives and economic efficiency provides a powerful source of agricultural growth.

Prior to 1977 such reasoning seemed to represent little more than a conjectural exercise. Since 1979 most of these suggestions have indeed been implemented on a widespread basis. By 1983, the responsibility system was practiced in some form by 98 percent of China's production teams and 10 percent of households have become specialized in a particular line of production (Zhiming 1983). The results appear to be astonishing. Data released by the State Statistical Bureau report that it took only five years to raise China's average per hectare grain yield by 870 kilograms (China Daily, May 3, 1984). In the interim period 1949 to 1965 per hectare grain yield went up by only 600 kilograms. Downgrading the command communal agricultural system by replacing it with a peasant household-based contract system and raising grain procurement prices to producers has spurred agricultural production to increase even as total sown acreage has declined. When officially the target to expand grain production to 400 million tons by 1985 was announced many were skeptical but grain output in 1984 is estimated in fact to have reached that target after six bumper harvests in succession (China Daily, August 25, 1984). So the question is no longer whether such reforms could lead to a surge in agricultural growth, but rather whether these increases can be sustained.

A pessimistic view would caution that China's agricultural productivity will be rooted in some fundamental resource and technology constraints. Perhaps China has now captured most of the gains from land and economic reform and spread of traditional and new seed/fertilizer technology. Improved water control has been a major source of high and stable yields, but what next? While China plans to quintuple the output of its chemical fertilizer industry by the year 2000, domestic supply has not been able to keep pace with demand. The country has had to boost imports to nearly one billion dollars annually (China Trade Report, 1984). Domestic production is increasing enough that the level of imports is

growing signs of decline but the fertilizer constraint is not yet assured of being removed. Mechanization is now primarily labor-saving rather than yield-increasing for most of China's agricultural systems. Does China's historic success in breeding modern, high-yielding cereal varieties mean that only limited opportunities exist for borrowing seed technology from the international agricultural research institutes? In addition, the domestic scientific capacity may have been badly eroded during the Cultural Revolution, thus implying a significant lag before new research results begin to have an impact in the fields and average yields of China's grain crops.

Which view is correct? One can speculate, but time, of course, will be the ultimate judge. With the economic reforms implemented by Deng Xioping, China's hope for continued increased agricultural production rests heavily upon an improvement in labor and management efficiency. Group incentives and political inducements to raise agricultural productivity were used very extensively during the Cultural Revolution, and the marginal impact may well have been significantly negative. To reap gains from improved labor and management efficiency, broader individual and production team initiative in resource allocation and greater financial rewards to successful efforts to raise output have been introduced since 1979. Such incentives have become visible both as reward to the successful and as spur to the failed. The incomes that accrue to successful peasants have been used to improve their consumption standards of living. The incentives strategy explicitly recognized that "some peasants must get rich first" with the result that income disparities are becoming more pronounced. China's leaders are testing the potential of this strategy because high priority is given to rapid gains in agricultural production. Such gains may help to avoid other, even more difficult, choices influencing the impact of China's overall development stategey on the world food system and resulting pressures on availability of grain to China itself.

In hinging future sources of increasing agricultural productivity on higher food prices as incentives to rural areas for both short-run intensification of food production and longer-run investment in agricultural infrastructure, China faces a food policy dilemma common in other countries. How can farm prices be kept high to maintain incentives for rapid growth in food output while keeping food prices low for consumers, especially those poor consumers who have until now benefited so significantly from the guarantee of cheap basic foods? Many countries have grappled with these issues, and comparative experience suggests that there are five basic mechanisms for resolving the policy dilemma. The categories discussed below are more tidy than real world experience; most policies are complicated and fuzzy amalgams of reinforcing and contradictory program elements. The following five alternatives do not constitute the reality of Chinese policy but rather discuss the alternatives and associated problems available to Chinese authorities.

Technical Change

Most long-term improvements in average standards of living have been generated by technical change. In the food policy arena "high" and "low" food prices are relative terms. High prices for farmers are relative to costs of production. Technical change lowers the costs of production per unit of output (real grain prices have actually fallen in world markets over the past thirty years) while retaining the profitability of producing that output.

The source of technical change varies from crop to crop and sector to sector. China has the option to increase the overall rate of technical change by devoting more resources to research and development of new technology. General and basic research may have serendipitous payoff in new agricultural technology, but most new agricultural technology is developed by applied research devoted to a specific task or crop. China can influence the rate of development and adoption of such technology through two mechanisms, direct public budgetary allocations to the research and to its extension to farmers and financial incentives via higher agricultural prices to induce both more innovation and adoption.

Investment in agricultural education has increased. Plans call for China's agricultural universities to have twice the enrollment in 1990 as in 1982. Secondary schools of agricultural education are slated to increase 2.6 times during the same period. The government has seemed eager to obtain technical assistance from foreign sources. Research operates in what must be a more conducive climate than what existed during the Mao era and the Cultural Revolution in particular (FAS-USDA, 1982).

Higher prices play a role in the adoption of technology also. Hence a major long-run role of higher food prices may be to stimulate the very technical change that is needed to make them "low".

Keeping food costs low is also a relative process. The actual cost of the average diet in the United States is very high by the standards of most of the rest of the world, and even a Chinese diet would be expensive in the United States by Chinese income standards. But food costs in the United States are actually very low relative to average U.S. income. High absolute food prices which are low food prices relative to the high incomes in a rich society can still cause severe problems of hunger for the remaining citizens with very low incomes. In such cases welfare programs and food subsidy schemes are essential to avoid significant hunger and malnutrition amidst the plenty.

If no mechanisms exist for protecting the food intake levels of poor consumers while higher food prices induce rapid technical change, then the food policy debate becomes focused on those short-run food consumption problems. China's food distribution mechanisms in both urban and rural areas may be subject to pressure if grain prices are to continue to rise significantly. In rural areas that do not respond to higher grain prices with higher grain production, strong pressures may exist to sell a larger proportion of output in order to capture the income effects of the higher prices. Low-income grain recipients may get caught in this squeeze.

Because of pressures on the government's budget the ration price in urban areas will eventually have to increase if the farm price is raised. With financial resources scarce and opportunities available for highly productive investment of budget revenues, urban consumers too may be squeezed by higher agricultural prices to producers. Low-income consumers might be protected by removing upper-income households from the urban rationing system altogether, but this would require an alternative source of grain for urban requirements. No matter how the dilemma is approached, higher prices as a stimulus to more rapid technical change create the need to design and implement short-run programs and policies that protect the food intake levels of the poor and yet are not antithetical to the long-run growth strategy itself based on technical change.

Subsidized Farm Inputs

For any level of output price, the incentive to increase production by more intensive use of an input can be improved by providing a subsidy to the cost of the input. Subsidized fertilizer costs are an especially common technique for increasing the profitability of intensive agriculture while keeping farm prices low. When total fertilizer use is low and incremental grain yield to fertilizer application rates are high, such subsidies can be a highly cost-effective strategy relative to output price increases or increased imports with subsidies. As fertilizer use becomes much more common, however, the costs of the program rise dramatically, and the production impact per unit of subsidy drops.

Other inputs can also be subsidized. Subsidized credit can also be used to encourage purchase and use of modern inputs. But none of the input subsidy programs are able to encourage the farmer to use more labor and

provide better managerial care for the crops. All subsidies tend to distort the intensity of use of inputs
from their economically optimal levels, thus causing
significant waste, unless significant underutilization
of the input exists due to imperfect knowledge and
greater risk aversion than is appropriate for society as
a whole. Since not all inputs can be equally subsidized, output price increases will have a greater productivity impact than input subsides, especially in the
long run. Consequently, input subsidies are probably
effective in keeping farm profitability high and consumer price low only for particular stages of input use
and for short periods of time. In addition, even the
short-run distortions may impede the efficient long-run
growth strategy.

Subsidizing the Gap Between Farm Prices
and Retail Prices

Subsidized marketing costs. China is not unique in
having the government absorb the real economic costs of
transforming the farm product in place, time, and form.
In various countries the government subsidizes the costs
of marketing grain either by establishing a government
agency with a budget provided from general tax revenues
to carry out the tasks itself or by subsidizing private
marketing agents to do the task. Since efficient mechanisms for subsidizing private marketing agents are
difficult to implement, most governments that wish to
squeeze the marketing margin to less than the full economic cost, including profits and return to entrepre-
neurial ability, end up doing the task themselves. Few
governments have demonstrated a real proficiency at this
task although some seem relatively adept, including
China. But such success, while lowering the cost to
those incurred by an efficient private marketing system,
does not thereby eliminate those marketing costs. They
are simply being paid by the state treasury and this
draws funds and resources away from other highly
critical needs.
The long-run efficacy of subsidized marketing costs
is difficult to judge. If future productivity gains in
Chinese agriculture are likely to come primarily through
more efficient resource allocation led by price planning
based on real opportunity costs, some better form of
receiving and transmitting signals than the state food
logistics agency is likely to be needed. There is evidence that the state marketing system has not responded
either efficiently or enthusiastically to the limited
experiments with incentives and greater peasant discretion in resource allocation, although freer rural markets may help.

Dual price systems. A number of countries, especially in South Asia, have experimented with dual price systems for basic foodgrains. Although program details vary considerably, the logic of the approach for a closed economy, i.e. with no food imports, is as follows: farmers pay a grain tax based on land cultivated or historic yields but not on current output. Thus farmers treat the grain tax as a fixed cost of production which does not alter their resource allocation decisions or short-run incentive to produce. The grain obtained from this tax is made available in government-operated or licensed "fair price" shops where low-income consumers are permitted to buy a ration quantity at very low prices. Farmers are free to sell their surplus production in an open market where consumers, including the poor, are free to buy whatever quantities they wish at whatever the market clearing price is. Hence there are two food prices in the system, the cheap ration price set by the government in the fair price shops and the free market price set by the equilibrium of supply from farmers and demand from consumers. In some systems the farmers are also paid a low "procurement price" for their grain taken by the state, thus reducing but not eliminating the tax element in the transfer.

A successful dual price system has several key ingredients. First, it requires access to significant quantities of low-cost grain from farmers large enough to produce a sizeable marketable surplus. Secondly, it requires careful control over access to the cheap grain available in the fair price shop. For the system to work, the ration quantities must be limited to the amounts available, and the rations must be restricted to the bottom end of the income distribution. If some of the poor are excluded from the system, they are doubly hurt, for not only do they not get the cheap grain from the fair price shop but the free market price is now substantially higher than it would have been in the absence of the dual price system. Hence targeted procurement and carefully targeted rationed sales are essential to the success of such a system.

Again, although it may appear that the system can operate without a subsidy, especially if the procurement price is sufficiently low (even zero) so that revenue from ration sales pays for procurement and distribution costs, the system clearly requires resource transfers from farmers to consumers. If the incidence of the transfers is on large wealthy farmers to benefit low-income consumers, the society may find this socially acceptable but this option does not apply to China. Given the absence of a wealthy farmer class for the present it is more likely the incidence would fall primarily on middle- to low-income farmers to the benefit of relatively high-income urban consumers, with the poor frequently excluded from the benefits. Also, a carefully designed procurement program can minimize the disincentive effects on agricultural production, but most programs have not been either designed or implemented so carefully. A common outcome is that prices for all farm outputs are depressed through the procurement program with sharply diminished incentives. Even for the most carefully designed program the tax equivalent of the grain procurement reduces saving resources available for private, farm-level investment in raising agricultural productivity. Such farm level investment typically has a fairly high payoff.

Subsidized Poor People's Foods

By economic necessity the poor in most societies eat different foods from those consumed by middle- and upper-income groups. Even in those countries where 70-90 percent of calories come from starchy staples, the diets of the poor are remarkably different from what an average food balance sheet would indicate. Poor people's foods tend to be rootcrops (cassavas, yams, sweet potatoes, Irish potatoes) or coarse grains (maize, sorghum, millet and others). The preferred starchy staple in most societies is either rice or wheat although corn is preferred in some African and Latin American societies. In rice cultures wheat is sometimes regarded as an inferior good.

Such sharp contrasts in food consumption patterns by income class within a country are not caused by differences in taste but by economic necessity. The poor in Indonesia eating cassava and maize would prefer to eat rice, as would the barley eaters in South Korea when the government encouraged substitution of barley for rice. If a society does not have the bureaucratic and financial resources to provide subsidies for the more expensive preferred foods, subsidies to poor people's foods may be effectively self-targeting. Since only the poor choose to eat the subsidized inferior staples, only the poor capture the subsidy.

Such subsidies have both short-run and long-run costs. In the short run implementing subsidies for commodities which may not travel or store well (rootcrops) or for which well-developed marketing systems do not exist may simply not be feasible without significant investment in food technology and improved marketing infrastructure. Because farmers tend to switch to growing more profitable crops, obtaining supplies of these commodities when market prices are being forced down is an obvious problem. Subsidies to farmers and/or to consumers become difficult to implement. Subsidizing an imported commodity with the desired characteristics may

be more feasible and have less impact on domestic
farmers. A subsidized low-quality wheat flour might
have this effect in Sri Lanka, for example.

The longer-run effects are perhaps more troubling
and suggest that disaggregated commodity price policies
probably can serve only as short-run bridges across the
basic food policy dilemma. The distortions introduced
by significant subsidies on a single commodity can eventually be very powerful. Livestock industries find
heavily subsidized corn or wheat a cheap, high-quality
animal feed. Such grain-fed livestock industries invert
the incidence of the subsidy from poor people to rich
people. The low prices for these inferior foods almost
inevitably reduce incentives for research and development of new technology for the crops and the profitability at the farm level of growing them. Consequently,
subsidies on inferior goods may be more cost-effective
at increasing calorie intake among the poor than subsidies on the preferred staple, but neither subsidy policy
is likely to be feasible in the long run because of
financial costs and impaired incentives to increase
agricultural output. Despite these problems, commodity
targeting of subsidies, implemented with a short time
horizon, may be easier and cheaper for some societies
than any other form of income-related targeting.

An Import-Export Strategy Using
Comparative Advantage

The favorite strategy of economists for dealing with
the basic food price policy dilemma is to import cheap
food from the international marketplace and export high-
valued cash crops and industrial goods in exchange.
China may look to Taiwan with envy as a country that
pursued this strategy with success. Taiwan pursued an
export promotion and agricultural development strategy
simultaneously. Agricultural production grew rapidly,
but the ironic twist is that now agricultural production
is a rapidly declining portion of the island's total
output and Taiwan is now a net importer of agricultural
products. The strategy can encounter difficulties.

China's success will hinge upon its ability to
export to earn foreign exchange. Textiles, shoes, and
light industrial goods exports have recorded notable
success stories but international protectionism could
restrict China's export drive in the future. The other
major source of foreign exchange that the Chinese hope
to tap is petroleum exports but a weakened market plus
disappointments in initial searches for oil in the South
China Sea have muffled the noisy optimism expressed in
the late 1970s for this source of bounty. Also imports
of technology, capital equipment and raw materials will

continue to increase as a part of the modernization drive (Adams 1984, p. 99).

Additionally, the grain from international markets may not be cheaper than domestically produced grain. Then a domestic subsidy is needed to make it "cheap". The logic behind such a strategy requires that domestic farmers would not respond sufficiently to higher prices to make the burden on consumers bearable. Consequently, the added supplies needed to keep food prices low are imported.

Another problem is that international grain markets are quite unstable. Even if an import-export strategy makes sense on average, it may not be successful in precisely those years when availability of imported grain is most critical, i.e. when domestic food prices are rising because of a short domestic crop. If this happens when international supplies are short, the reliance on the international market to make comparative advantage a source of dynamic efficiency growth may be risky indeed. Taiwan had P.L. 480 and other sources of external aid that it could rely on but the People's Republic may have less assistance and luck. Planning authorities can be expected to factor the psychological risks of international price instability into their calculations of the costs of importing grain (Jabara and Thompson, 1979).

Nevertheless China indeed seems to have opted for a trade strategy as a part of its food policy. Turning to the international market for significant quantities of grain (10-17 mmt a year is significant) is a major reversal from the policy of self-reliance espoused by Chairman Mao. Most economists have applauded the new openness of the Chinese economy as a healthy step toward increased specialization and improved efficiency, although many agricultural economists will be watching closely to see if food grain imports are used to depress farm incentives. But an essential ingredient of using the international grain markets successfully, whatever the motivation, will be an improved understanding of how those markets are likely to work in the 1980s and beyond relative to their behavior in the 1960s and 1970s.

UNDERSTANDING WORLD GRAIN MARKETS

The decade of the 1970s marked a transition from the steady trends toward new patterns of international grain trade that had emerged in the 1950s and 1960s to a world grain economy characterized by severe instability in trade patterns and in prices and dependent on North America as the major net surplus grain producing continent. The instability was partially the result of political decisions in the Soviet Union to use world grain supplies to smooth out its inherent fluctuations in basic grain production and partly the result of the U.S. political decision not to hold large grain stocks, which in previous decades had the important, if unintended, effect of stabilizing world grain prices.

The emergence of North America as the residual grain supplier to the rest of the world has been gradual since the 1930s, but the trend accelerated in the 1970s as the dollar was devalued and as corn became the fastest growing grain in world trade. Corn's trade patterns correspond closely to the patterns of rapid growth in incomes. Most of the increase in corn imports has been in Japan, South Korea, and Taiwan; in Eastern Europe and the Soviet Union; in the faster-growing countries of Western Europe; and in a few other rapidly growing, middle-income countries, such as Mexico, Brazil, and several OPEC nations. Since the United States supplies three-quarters of the world's exportable corn and since corn has become a much larger proportion of world trade (from roughly an eighth in 1960 to well over a third in 1980), the world has come to rely much more heavily on the success of North America's grain crops to provide readily available supplies at stable prices. When bad weather hits the Corn Belt, the entire world watches nervously.

The drive to find cheap sources of livestock feed could provide much of the momentum to international grain markets in the future as it did in the 1970s. Most of the world's corn, and nearly all of the corn that enters international trade, is fed to livestock. Over the past two decades corn has emerged as the lowest cost source of a nutritionally well-balanced feedstuff, combining carbohydrate, fat, and protein in a nearly ideal combination for several important categories of livestock. Sophisticated livestock feed blends can supplement and improve on a corn base, but corn is the base for much of the emerging feedlot meat production around the world. Understanding the world's corn economy in the future may be a key to understanding the rest of the world's food system.

What happens over the next decade to prices for basic food commodites depends on supply factors and demand factors. As the world economy emerges from the energy and price induced recession of the early 1980s, income-led demand for meat and its resulting ramifications on demand for livestock feed relative to the supply of the latter could re-emerge as an important force pushing the world grain market. This scenario of course hinges upon whether and to what extent economic growth occurs. Under a growth scenario strong leverage is created by the inefficiency of converting grain to meat; roughly ten grain calories are required to produce one calorie of edible meat using intensive feedlot technology. Lester Brown has calculated that a single addition to the world's population consumes about 180 kg of grain per year if it is eaten directly as food as a subsistence diet. But if that individual is combined with adequate income to consume an affluent, meat-intensive diet, then the grain demand rises to roughly 730 kg per year.

For China, given the numbers of people it feeds, it is important to watch income growth and population growth. Population growth will continue to stimulate demand for grain through the year 2000 even though the rate of growth is targeted to decline. In 1970 the natural annual rate of growth was 2.6 percent but the State's population control program (one child -- one family) brought this down to 1.17 percent in 1979. Population is targeted to be 1.2 billion in the year 2000. This implies an annual growth rate of 1.0 percent between 1980, when population was estimated at 982.6 million, and the year 2000. Whether China will succeed in curtailing growth to this level remains to be seen. Indeed the rate in 1980 was 1.21 percent, up from the preceding year. Part of the explanation is that the demographical composition of China's population is becoming distributed such that the proportion of people in the child bearing age is increasing. Moreover the rural responsibility system has tied family income more closely to work performed by the family unit, thus increasing the incentive to have more children. Therefore, in spite of the admonishments and disincentive programs designed by authorities to restrict each family to one child, it is questionable that population increase can be kept at the one percent rate. Even if it is, total population will increase by 22 percent in a twenty year period adding nearly twice as many mouths to feed as the current population of Japan or the total population of the United States.

Population growth adds to food needs slowly and steadily. However, not all of the added food needed will show up as effective demand in world grain markets. An inherent characteristic of income growth, however, is that it can be converted directly to market demands. Income growth in societies above $500-600 per capita per year tends to cause extremely rapid increases in meat and livestock product demand. Despite much of the world's attention to the failures of economic development in some countries and for the very poor in many countries, a rapidly growing proportion of the world's population has in fact emerged from subsistence level poverty and is clamoring for improved standards of living, especially through a better quality diet with more meat. Since more meat at the margin means grain fed to livestock, the emergence of a large fraction (China in particular) of the world's population into the relatively affluent middle class could place increasing demands on world supplies of livestock feeds. Income growth especially in the middle-income countries, has become the major driving force in world food grain markets. China is not likely to achieve such status until later in this century, but if and when this occurs, given its size, food policies there could have consequences for the world food system reminiscent of the role of the Soviet Union in the 1970s.

Substitutions among commodities in end uses will then transmit much of that driving force to other commodities that are strongly linked within the system itself.

CHINA'S USE OF WORLD GRAIN MARKETS

Understanding how and why China uses world grain markets for its own domestic purposes is a topic for China specialists who are able to follow both the trade trends and policy environment. This section is designed simply to lay out the general historic framework in which China has participated in the world food economy as a precursor to turning the direction of the question around. Then the issue will be the impact of China on the world food system and any feedback effects on China itself.

China's involvement in world grain trade seems to correspond with periods when economic growth and efficiency are receiving high priority in Beijing. This is no accident. Trade is the only way to capture the gains of specialization, and specialization through comparative advantage is the basis for high productivity and improved standards of living. This has been true at least since Adam Smith discussed his famous pin factory, and it remains true today. The difficulty is that the gains to specialization and trade seldom accrue equitably to all members of society, especially in the early days of development. Consequently, many countries have attempted to inhibit specialization and trade in an effort to preserve or create a more equitable distribution of income. Unfortunately, the policies that do this are usually inimical to economic growth. As growing population begins to put pressures on living standards, especially food consumption levels, the necessity for economic growth becomes more and more apparent. Hence countries often go through policy cycles alternately favoring and discriminating against international trade. It is not surprising that China has repeatedly faced these pressures as well.

Because of its ability to alter somewhat the relative consumption of wheat and rice through the urban rationing system, China has seemingly used world grain markets to carry out a calorie arbitrage. When rice prices are very high and supplies are especially tight, China moves into the rice export market. Because of its

low quality of export rice, China is unable to compete
with the United States and Thailand on favorable price
terms during weak markets; but in times of scarcity,
Chinese rice finds ready buyers. At the same time China
replaces the exported rice on a ton-for-ton basis with
imported wheat, usually at about half the cost. In
recent years China's wheat imports have considerably
exceeded the dollar value of its rice exports; but there
were years when a rough balance was struck with little
impact on China's overall trade balance, and with considerable gain to China's consumers (unless they had a
strong preference for rice).

World grain markets also benefited from this trade
because the world rice market is very thin, only about
three percent of production is traded internationally,
while the world wheat market is much larger and able to
absorb added demand of two to three million metric tons
without great price gyrations. By adding to world rice
supplies at peak price periods (and withdrawing while
the market is weak), China has served as a very considerable price stabilizer. Since very few major rice consuming countries want (or if possible, allow) changes in
world rice prices to affect their domestic consumers,
China's ability and willingness to absorb some of the
adjustments needed to permit the world rice market to
clear is extremely important.

China's actions in world grain markets suggest it is
aware of the potential to use international prices as an
opportunity cost standard against which it might judge
domestic policy alternatives. Clearly the decision, if
it was intentional, to import cheap wheat and to export
expensive rice, rather than consuming expensive rice and
attempting to produce more expensive wheat, reflects
this perception. More recently the increase of cotton
production (an important input in China's rapidly growing textile exports) on acreage previously devoted
primarily to grain, testifies to awareness of this
strategy. How far China will go with this strategy is
an important question. Comparing the domestic resource
cost of planned production (and consumption) efforts
relative to the availability and cost of foreign
exchange needed to import the equivalent output can be a
key planning mechanism to promote efficient growth,
whether in a socialist or a market economy. Such an
opportunity cost perspective is only one ingredient in
the planning process, of course, and even it must be
used with considerable flexibility to avoid mechanical
answers in the face of a highly uncertain and unstable
international economy. In addition, for China the
opportunity cost perspective becomes even more complicated because of China's potential to influence the international terms of trade by its own actions. China's
actions in the world grain economy in particular could
be sufficiently influential to cause major readjustments

to current expectations about the likely path for major commodity prices over the next decade.

CHINA'S POTENTIAL IMPACT ON WORLD GRAIN MARKETS

China's recent grain import levels of up to 17 mmt were large enough to have a significant impact on grain markets and price levels. China's share of the world grain market in the post Mao period is approximately 15 percent as compared to 5 percent in 1960. The per capita import amounts are now roughly equivalent to those of India during the worst of the harvest failures in 1966 and 1967. Those imports produced considerable anxiety about whether the world could feed itself and the potential necessity to "write off" India as a hopeless case. Ironically, the equivalent import levels in China have gone relatively unnoticed by the world press and are treated as a healthy sign by many informal observers. Part of the difference is that China has for the most part paid cash for the grain while India summoned world attention to her plight and received massive food aid shipments from the United States with considerable political conditions attached. Also supplies of grain in the world market are relatively abundant now.

In a recent Cornell University study (Schwartz and Ralston 1983) the potential impact that Chinese wheat imports could have upon world prices and trade during the remainder of the 1980s has been analyzed by means of stochastic simulations of an econometric model of the world wheat trade. Projections of Chinese import levels though 1989 were based upon three scenarios developed from different growth rate assumptions relating to grain production and population and consumption requirements in the rural and urban sectors.

China specialists frequently note that Chinese wheat imports have been used to meet the needs of coastal urban deficit regions. Since rural consumption relative to production affects the availability of domestic wheat supplies to meet these urban needs, focus upon these sectors and their interaction seems especially appropriate. The projections also recognized the significance of the distribution of population by age.

Of the three scenarios the one with the least impact on imports assumed government policies could successfully restrict population and consumption growth along with a return to conservative policies typifying the era from 1965 to 1976. Under this scenario imports could fall below their current levels. The other two scenarios assumed either moderate tightening of central control over production and population under which imports are projected to reach 17 mmt by 1989, or a continuation of events that open the Chinese economy and upgrade diets in which case imports could climb to 22 mmt. This range of import possibilities is projected in the study to comprise from 13 percent to 23 percent of total wheat trade. This potential range was estimated to affect the expected value of world trade by over 40 percent and more than triple relative price variability due to stochastic variations in world wheat trade in conjunction with the high import scenario. The authors conclude, as we do, that it is essential that policymakers need to monitor China's direction with regard to population, income, and foreign exchange options since "the price disruptions in the world wheat market experienced in the mid 1970s could be reexperienced unless precautions are taken" (p. 27).

China now buys two major commodities from international grain markets, wheat and corn. Although wheat is presently much the more important in both volume and value terms, China's major potential impact on the world food system could be through the corn market. The two markets are linked in two ways: (1) about one-third of the world's wheat is fed to livestock; so if corn prices rise to roughly 90 percent of wheat prices, then the two grains are linked as equivalent foodstuffs; and (2) one possible development strategy for China that implies significant imports of corn to expand meat production rapidly may have as a side effect the reduction in wheat production and hence lead to significant increases in wheat imports as well.

All discussion of Chinese grain imports must be done in the context of foreign exchange availability and alternative uses. China's reserve holdings were estimated to exceed $16 billion by 1984, which was up from $2.3 billion at the end of 1980 (Bogert 1984, p. 39). Recently Chinas' foreign exchange reserves have been depleted by a buying spree. In the six months to March 31, 1985 these reserves dwindled by 30 percent from 16.3 billion dollars to 11.3 billion dollars (Journal of Commerce, July 16, 1985). Long-term availability of foreign exchange will undoubtedly continue to constrain imports. All of the high Chinese import figures to be discussed below are unlikely to be realized precisely because of foreign exchange constraints, even if the Chinese wished to increase dramatically their foreign debt, which is not probable. But this is precisely the point of the discussion. If the import levels "required" for a given strategy to work are not possible, then the strategy itself must be altered and modified. Identifying the forces pushing toward an equilibrium is the task here.

Consider a "high growth" economic development strategy for China that is based on the use of material incentives to spur urban and rural workers to higher productivity and greater efficiency in using inputs to produce output. A strong consensus seems to exist among outside China scholars that such an incentives-led strategy is probably the only path that can dramatically alter China's historic growth path and lead to real progress in meeting the goal of a rich and powerful China by the year 2000.

Some of the implications of such a strategy for agriculture have already been discussed in the context of the food price policy dilemma created by high agricultural prices relative to low-income consumers' purchasing power. The possible food consumption consequences of an incentives-led growth path are less troubling for China, with its urban and rural food grain rationing systems in place and working effectively, than for market-oriented economies without such guaranteed minimum consumption mechanisms. But even for China the costs of the guarantee rise significantly and cut into state revenues for investment in productive plants and infrastructure.

One side effect of a freer rural decision making environment and added incentives to produce industrial and cash crops according to comparative advantage is likely to be a reduction in wheat and rice output, or at least a reduction in their rate of growth. As more energy and resources are devoted to cotton, sugar, fruits, vegetables and animal husbandry, less will be available for basic grains. Since they will be relatively less valued by society, judging from the prices paid to farmers, this will make social sense as well, at least in the short-run. But the pattern of long-run equilibrium is not so clear. If grain output fails to keep pace with growing demand then imports will be needed to maintain the desired policy-determined price. Even if such prices are in relative parity to world prices, substantially larger wheat imports will strain foreign exchange availability (not to mention port facilities) and probably the exchange rate itself. A market adjustment mechanism would then force up domestic grain prices, thus encouraging peasants to grow more and consumers to eat less. Since low-income consumers are most sensitive to such price changes, their nutritional status will deteriorate in the absence of preventive interventions to protect them from lowered food intake. How might China deal with such a dilemma? One obvious possibility is by not freeing resource allocation in rural areas quite so dramatically and by not changing relative profitability of cash and industrial crops versus grain. Much growth potential may be foregone as a result, but the wheat import bill will remain manageable and low-income consumers will be protected, at least for a while.

A second major aspect of an incentives-led strategy focuses on the impact of new consumer demand patterns generated by the higher incomes that result from greater production and productivity. The left hand side of Chart 6.1, with diamond shaped boxes, illustrates the

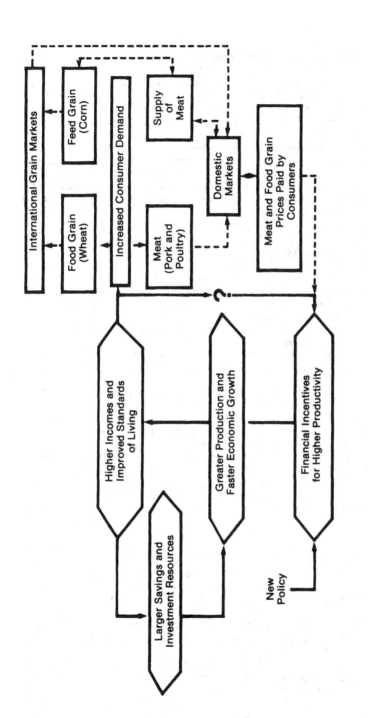

Chart 6.1   The Virtuous Circle of Incentives and Higher Productivity

"virtuous circle" of incentives, higher productivity, higher incomes and greater incentives that is the basis of any incentives-led growth stragegy (including President Reagan's in the United States). The circle is started on its upward spiral by a policy intervention that provides workers with personal incentives to increase productivity. These incentives lead to greater production and more rapid economic growth which permits higher worker incomes and the potential for higher standards of living. The higher incomes have two virtues: some of the increase is saved and devoted to investment in new plant and equipment, thus raising productivity potential even more. The rest feeds back to reward harder work through a second round of higher incentives.

Although there are several places where such a virtuous circle can run off course, the important one for the present discussion is illustrated on the right hand side of Chart 6.1, with dotted lines connecting rectangular boxes. This right hand sub-system is connected to the main economy and its productivity growth by the spending of higher incomes on goods and services desired by the working population. Higher incomes create increased consumer demand. In the Chinese context some of this increased demand will be for foodgrain directly, especially wheat, and much of this will spill over directly into greater imports from international grain markets.

An even more powerful effect on grain markets could result if China upgrades consumption of meat and livestock products. China's meat consumption per capita, mostly pork and poultry, is among the lowest in the world and is exceptionally low for its income and calorie intake level. Even granting a strong traditional taste for wheat and rice, there is a very large potential and pent-up demand for meat among the Chinese population. In the context of an incentives-led economic growth strategy some of this latent demand must be permitted to express itself through greater supplies of meat available in the market. A recent press statement claimed that average consumption of pork which is China's basic meat has increased to 12.3 kilograms per capita or 4.65 kilograms above the level in 1978 (China Daily August 21, 1984 p. 1). Total red meat availlibility per capita which also includes beef and mutton increased from 8.9 to 13.7 kilograms per capita (USDA). If meat consumption continues to increase at this pace China's domestic growth and/or trade policy will feel the effect. If meat prices rise sharply because of short supplies, then much of the intended incentives to produce will be eroded, thus dampening the growth effort itself. Increasing meat supplies rapidly is likely to be a key ingredient to a successful incentives-led development strategy in any of the middle-income countries and especially in China. While pork will remain the mainstay of this trend, the poultry and dairy sectors will also demand added feed inputs. Vigorous promotion of modern dairy production has been underway with milk output increasing from 971,392 tons in 1978 to 2,295,000 tons in 1983.

What will be the impact of such an effort? Supplies of pork and poultry to date have been primarily increased through traditional animal husbandry techniques but scavenging as a feed source is already intensively used and rapid growth in meat production at some point will come only from a significantly improved diet for China's pigs and chickens. This has in fact already begun to occur. Production of compound feed has quadrupled from 1.10 million tons in 1980 to 4.5 million tons in 1983 (USDA). Fully mechanized feedlots probably require about ten kilograms of grain per kilogram of consumable meat, but marginal weight gain on existing animals raised in a traditional manner probably requires only three kilograms of grain per kilogram of gain.

Assume that Chinese per capita incomes increase at 5 percent per year from 1980 to 1990, a conservative figure if the growth strategy gathers steam. If income elasticities for meat are about 1.0, a figure consistent with other populations at this stage of development but quite low relative to present levels of meat consumption, then meat demand will also grow at 5 percent per year if prices are stable. If this demand is allowed to be expressed in the marketplace, meat supplies must increase about 500,000 tons per year to keep pace. If all of the meat is produced in feedlots, then as much as 5 mmt of added grain per year will be needed. If it is all imported, China could be importing 50 mmt of corn or other feedstuffs by 1990. If all the meat was produced at the margin from traditionally raised animals fed a grain-intensive diet for more rapid gain, only 15-20 mmt would be needed by 1990. But the logistical difficulties of actually getting all this feedgrain to traditional livestock producers are enormous. The government is emphasizing rapid development of cattle and sheep industries in the vast grazing areas of inner China (Adams 1984). Some combination of the two techniques might be more likely.

Barring complete success with the program to expand meat output from cattle and sheep fed on grass, it is plausible to see China adding 20-30 mmt in corn import demand to the world market by the end of the century. If that market should be under severe demand pressures and tight supplies, such pressures would surely feed up to the world wheat market as well and thus raise the costs of wheat imports. Since each additional million metric ton of corn at four dollars per bushel ($160 per metric ton) delivered in China costs $160 million, the foreign exchange burden of such an outcome is obvious and almost certainly would be unacceptable. Consequently, conditions in world grain markets if China actively pursues a vigorous incentives-led growth strategy could tend to send cautioning signals back to the Chinese economy. In the absence of a resilient world grain market with substantial excess capacity, China's development strategy may fail because of its success. The alternative, vigorous and dynamic growth from China's agricultural sector itself, is filled with equally troubling dilemmas and internal inconsistencies if the resource and technology base is as close to potential as some observers think.

In summary the balance of the twentieth century is likely to be one of significant trial and error as the Chinese economy seeks a feasible, equitable, and efficient path to higher standards of living. Frictions and market disruptions will appear with increasing frequency if China increases its imports of grain and the impact will feed both forward to the world food system and back to the Chinese economy. In short China is an enigma in the world grain trade that cannot be ignored.

A NOTE ON SOURCES

Although few references or footnotes are used in the body of the paper, it is obvious that heavy reliance has been placed on the work of others in preparing this perspective on China and the world food system. Much of the material used is from unpublished drafts that are not available for citation; a manuscript by Kathleen Hartford falls in this category. Papers by Nicholas Lardy and E.R. Lim, for example, were extremely helpful as was the World Bank study (1981).

The influence of Dwight Perkins on the perception of what Chinese agriculture has already accomplished and the potential for future growth is clear throughout the paper. Although we cannot claim to have a detailed knowledge of Chinese agriculture, we can claim to agree with someone who does. This does not implicate Dwight Perkins in the views presented here, but if he is significantly wrong in his assessment of the resource constraints facing Chinese agriculture, then most of the dilemmas discussed in this paper will be softened.

Similarly, a substantial debt is owed to Nicholas Eberstadt's "Poverty in China". His review rightly flagged the article by Timmer, Falcon and Nelson, on "China's Food Policy" as entirely too optimistic on the degree to which China had eliminated basic hunger. Eberstadt's integrative analysis is persuasive, and much of his perspective is reflected in this paper. Nicholas Lardy has recently added further insight on China's food problems.

All China's scholars (and especially non-specialists such as ourselves) are indebted to IFPRI (International

Food Policy Research Institute) for commissioning the study by Anthony Tang and Bruce Stone of "Food Production in the People's Republic of China." The data collected and the analysis performed now serve as a basis for all future work.

Another especially useful source of data and current outlook is in the United States Department of Agriculture's annual review, "World Food Situation: the People's Republic of China". Charles Liu and Fred Surls, in particular, have provided that document with analytical perspective not often found in outlook reports. Also Fred Surls generously provided constructive comments on an earlier draft of this chapter that were extremely helpful.

Lastly, recognition is due to a very stimulating conference on Chinese agriculture sponsored by Cornell University a few years ago. The participants represented a very wide spectrum of opinion, and the discussion was sharp and useful. Randy Barker and Radha Sinha produced a summary of this conference which captures much of that spirit, "Cornell Workshop on Agricultural and Rural Development in the People's Republic of China."

NOTES

1. We are indebted to Fred Surls for correcting our initial impression that urban prices were below rural prices prior to 1979.

REFERENCES

Adams, Mike. "Chinese Agriculture and Agricultural Trade: An Evaluation to the Mid 1980's". in Adams, M.G. (ed) Economic Development in East and South-East Asia: Implications for Australian Agriculture in the 1980s. AGPS Canberra (1984): 87-105.

Bogert, Carrol R. "Americas Open Door". The China Business Review. Volume 11, No. 5 (October 1984): 39-43.

China Daily. "Annual Grain Output Hits Record High." (May 3, 1984): 3.

China Daily. "Grain and Cotton Output on Target for All-time High." (August 25, 1984): 1.

China Daily. "Bound for Table." (August 21, 1984): 1.

China Trade Report. "Nurturing the Barren Land." (January 1984): 8.

Economic Research Service, USDA, China Outlook and Situation Report. RS-89-8 (June 1984).

Foreign Agricultural Service, U.S.D.A. Foreign Agriculture. (August 1982): 13.

Hsu, Robert C. Food for One Billion: China's Agriculture Since 1949. Boulder, Colorado: Westview Press (1982).

Jabara, Cathy and Robert Thompson. "Agricultural Comparative Advantage under International Price Uncertainty: The Case of Senegal". American Journal of Agricultural Economics. (May 1980): 188-198.

Journal of Commerce. "Chinese Effect New Import Taxes." (July 16, 1985):p. 3A.

Lardy, Nicholas R. "Mao's Great Agricultural Leap Backward." The Asian Wall Street Journal. July 16, 1984): 6.

Schwartz, Nancy E. and Katherine Ralston. The Impact of Chinese Wheat Imports On World Price and Trade. Cornell University Agricultural Experiment Station, A.E. Res 83-28 (June 1983).

Zhiming, Jiang. "Sustained Growth of China's Economy in 1983." China Market, No. 11 (1983).

7

Agricultural Trade Implications of COMECON 1981-1985 Plans

Stephen C. Schmidt

INTRODUCTION

In analyzing the prospects for agricultural imports in the centrally planned economies it is important to look at their five-year plans. These plans indicate general policy emphasis and objectives. Of course five-year plans and actuality do not always conform, so it is important to continually evaluate official targets in terms of past experience and the likelihood of future events conforming with the scenarios that serve as the basis for these targets. Official 1981-1985 plans of the Council of Economic Mutual Assistance (COMECON) countries in terms of production and consumption intentions and the implications for agricultural trade during this period are reviewed. Projections by several independent sources of what grain and oilseed imports will be are also surveyed.

General Objectives and Aspirations

The targets for agricultural expansion in the 1981-1985 plans are largely clustered in the range of 2-3 percent rate of increase with slightly higher targets in Bulgaria (3.7-4.0 percent). These targets as well as past production results and targets are summarized in Table 7.1. Romania is again the exception with a very high planned rate of growth of 4.6-5.0 percent. The main policy emphasis in all countries of the region is on; 1) meeting the populations' demand for food not only quantitatively but also by raising food quality, 2) achieving more efficient use of resources, 3) adjusting the structure of production conforming with domestic and external demand developments, 4) reducing present import levels, and 5) stepping up exports of processed products. Basically all countries seek agricultural self-sufficiency or surplus production in order to reduce

TABLE 7.1
Agricultural Production Targets and Performance, 1961-1965 through 1981-1985.

Total & Average Annual Percentage changes		Bulgaria	Czech.	GDR	Hungary	Poland	Romania	Soviet Union
1961-1965								
Actual:	Total	23	2.5	9	8	14	17	15
	Annual	4.3	0.5	1.8	1.6	2.6	3.2	2.8
1966-1970								
Plan:	Total	25-30	15	13-15	13-15	14-15	26-32	25
	Annual	4.6-5.4	2.7-2.8	2.5-2.8	2.5-2.8	2.7-2.8	4.7-5.7	4.6
Actual:	Total	26	19	9	16	9	24	21
	Annual	4.8	3.5	1.9	2.8	2.9	4.2	3.9
1971-1975								
Plan:	Total	17-18	14-15	12	15-16	19-21	36-49	20-22
	Annual	3.2-3.7	2.6-2.8	2.4	2.8-3.0	3.5-3.9	6.8-8.3	3.7-4.0
Actual:	Total	12	16	11	18	17	26	13.3
	Annual	2.3	3.0	2.1	3.5	3.2	4.7	2.5
1976-1980								
Plan:	Total	20	14-15	14	16-18	16-19	28-44	14-17
	Annual	3.7	2.6-2.8	2.6	3.2-3.4	3.0-3.5	5.1-7.6	2.7-3.2
Actual:	Total	11	9	6	15	2.5	27	9
	Annual	2.1	1.7	1.1	2.9	0.5	4.9	1.7
1981-1985								
Plan:	Total	18	5.2	6	12-15	9[a]	25-28	12-14
	Annual	3.4	1.0	1.1	2.3-2.8	3[a]	4.5-5.0	2.3-2.7

[a]Refers to 1983-1985.

Source: National statistics of the socialist member countries of COMECON.

imports and expand exports. Agriculture continued to remain a priority feature of a generally restrictive investment policy.

To increase production each country adopted various types of incentive measures and took steps to promote increased industrialization of agriculture. The former include the raising of procurement prices and introduction of bonuses related to output deliveries. Improving the quality of life in rural areas by construction of additional housing, roads, schools and cultural and service establishments is also part of incentive programs.

Growing attention is being given to problems of rigidities in central planning and control because the system of plan indicators are no longer regarded as sufficient to ensure production efficiency. There is experimentation with relaxation of the central planning and control system with part of these functions being transferred to economic organizations and their associations. Most progress toward management decentralization, so far, was made in Hungary, Bulgaria and Yugoslavia. In recent years the German Democratic Republic (GDR) has started to loosen control of farm management. It remains to be seen how much authority Czechoslovakia, Romania and the Soviet Union, with still rigid centrally managed economies, will be willing to grant to their local farm managers over day-to-day operations.

Stress is being placed on cost accounting and financial autonomy. To this end measures were taken toward a more realistic alignment of retail and producer prices with production costs in an effort to reduce state subsidies. On the whole, however, sharply rising production costs due to higher energy and other raw material prices have not been fully offset by higher producer prices. Retail food prices too, were generally raised thoughout Eastern Europe to curb demand.

There is a growing awareness among policymakers of the decisive importance of private agriculture in meeting production targets in the current five-year plans. This recognition has led the governments to permit expansion of the private sector. The process by which this is being achieved varies by countries.

As usual, the broad trends and developments described for the region as a whole cover a considerable range of differences at the country level. Some salient features of individual country differences are highlighted in the following.

NATIONAL AGRICULTURAL PRODUCTION GOALS

Bulgaria

In the course of rapid industrialization, the relative importance of agriculture in the Bulgarian economy has fallen constantly, especially since 1965. Nonetheless, agriculture is still an important sector of the national economy, accounting for 16.7 percent of national income in 1983 and employing 23.3 percent of the total labor force.[1]

Agricultural production during the 1981-1985 period was planned to grow by 20-22 percent, or at an annual rate of 3.7-4.0 percent, representing a modest increase over the preceding quinquennium (Table 7.1). Livestock farming was to expand at a slightly lower pace, 17 percent or at 3.2 percent a year. Here the stress was to be placed on the development of cattle and sheep breeding, apparently with a view of generating exportable surpluses. Planned increases in pig and poultry production are primarily aimed at satisfying domestic needs.

Crop production plans. The food and feed grain sector is ascribed particular importance. Bulgaria's climatic and soil conditions enable a broad range of crops to be grown. Farms should increase grain production to produce 10.5-11.0 million tons annually by the end of the 1981-1985 period, indicating an increase of 35 percent compared with 1976-1980 (Appendix Table 7.1). In the preceding quinquennium Bulgaria was aiming for an average grain production of 9.3 to 9.6 million tons but produced only 7.8 million tons a year. The 1982 outturn of grains at 9.9 million tons, is the largest harvest to date. Grain production totaled 9.3 million tons in 1984.

Wheat is the major grain crop, accounting for 52 percent of total grain output in 1984, corn about 33 percent and barley 14 percent. Irrigation and increased fertilizer applications are seen as the main ways in which this increase can be achieved.

In regard to other crops, the area devoted to sugar beets, sunflower, soybeans, grapes and tobacco is to be increased. It was planned to grow soybeans on irrigated land only and increase its area to about 300,000 hectares by 1985 as compared with about 75,000 hectares in 1984. It was also decided to increase yields per hectare on grasslands to 3.5-four tons by 1985. On the basis of present yield levels this target is unlikely to be achieved. Additionally the plans called for increased fruit and vegetable production to meet domestic requirements and to meet the demand of specialization and cooperation among the COMECON countries.

Bulgaria has few reserves of agricultural land left. The main factor in the growth of crop production is expected to be improved yields. Agriculture during the preceding five-year plan received 13.4 percent of total investment against the planned 15.6 percent. Indications are that the rural sector had to be content with a smaller share of total investments.

Czechoslovakia

Agriculture occupies a relatively minor position in the national economy, contributing 8.4 percent to national income in 1982--one of the lowest proportions in Eastern Europe. The agricultural labor force represents 12.5 percent of the total labor force.

Czechoslovakia has the most diverse topographic area in Eastern Europe which has contributed to the development of a diversified agriculture in the country. Notwithstanding this diversity, Czechoslovakia's soil and climate are generally favorable and suitable for the cultivation of a wide range of crops.

The targets for 1981-1985 are more modest than those of the previous five-year period. The revised target for total farm production was a five-year rise of 5.2 percent against the previous quinquennium, about half the original plan (Table 7.1). An important objective of the plan was the gradual achievement of self-sufficiency in grain production and continued improvement in the level of self-sufficiency in the production of all foodstuffs. In line with these efforts, crop production received priority over animal husbandry in 1981- 1985. Crop output was to advance 10.8 percent total and livestock production by no more than 6.1 percent, thus increasing the share of crop production in the gross value of agricultural production. Capital investment in agriculture amounted to 20 percent of total investments in 1984.

Crop production plans. It was planned to achieve an average annual grain output of 11.2 to 11.6 million tons, 11 to 15 percent higher than the ten million ton average recorded during 1976-1980 (Appendix Table 7.1). Grain crops failed to meet their eleven million ton target in 1981 and 1982, but reached records of 11.04 million and twelve million tons in 1983 and 1984 respectively. Wheat and barley account for 82 percent of this production. The standing goal is near self-sufficiency in grain production for the medium term through 1985, and full selfsufficiency by 1990. Among grains, priority is to be given to corn, barley and oats, at some expense of wheat.

Domestic oilseed production has expanded sharply and reached about 350,000 tons in 1984. Of this, rapeseed accounted for 300,000 tons and sunflowerseed 44,000 tons. A national oilseed program got under way in 1983 to try to accelerate development, with the long-term aim of making the country self-sufficient. Presently domestic crops provide only about 30 percent of protein requirements.

Livestock production plans. Livestock production was to expand by a mere 6.1 percent compared to the 9.6

percent output gain in the previous quinquennium. Overall meat output was planned to increase by 11.2 percent,
to 1.58 million tons, and milk by 11 percent, to 6.2
million tons. Meat production in 1985 is expected to
reach its target level. The important new feature of
the 1981-1985 plan is an attempt to adjust livestock
herds to domestic feed supplies.

The structure of livestock production is being
altered through reduction of hog numbers, consolidation
of poultry numbers and enlargement of cattle and sheep
herds. Cattle numbers were to increase by 10 percent
over the 1980 figure of five million head. Hog numbers
fell to 6.75 million heads by January 1, 1985 in order
to bring inventories in line with the domestic feed
base. Investment for the food processing sector was
planned to be less than before and directed into existing plants for better efficiency. The plans stressed
increased production in the meat and dairy product processing industries. It also called for expanding the
range and quality of products, increasing production of
ready-to-serve and semi-prepared foods and improved food
packaging.

German Democratic Republic (GDR)

Agriculture is of a comparatively small importance
within the GDR economy. In 1983 agriculture and forestry contributed approximately 7.7 percent to national
income and employed 10.7 percent of the total labor
force. The share of agricultural investment of total
investment was 9.6 percent in 1981--reflecting a considerable influx of capital into this sector.

The 1981-85 plan aimed at the highest possible level
of self-sufficiency, thus reducing the country's dependence on western agricultural imports. Agricultural
production was projected to increase at an average
annual pace of 1.1 percent with emphasis on increasing
self-sufficiency in crop production, particularly grain
(Table 7.1). A related goal was to end the imbalance
between grain and protein production and livestock production that was aggravated over the past five years
when domestic feedgrains and protein feed were insufficient to meet requirements of larger herds.

The 1985 production target was set at 10.4 million
tons with a gradual increase to that amount in the
intervening years. This compares with the 9.5-10.5
million ton target set for the 1976-1980 plan period and
an actual output of nine million tons (Appendix Table
7.1). The 1985 output target was to be realized by
yield improvement and some expansion in planted areas.
The plan envisaged a yield of 3.95 tons per hectare by
1985 compared with 3.74 tons per hectare in 1980-1981.

The GDR attained a succession of record harvests since 1982 with production coming to 11.5 million tons in 1984. Yields too hit a record 4.49 tons per hectare in 1984. Higher producer prices seem to have contributed to larger output. The GDR has announced that its 1985 production target for grains - 10.7 million tons - has been fulfilled.

Regarding individual grains, a further increase in the area planted to baking quality wheat and barley is expected. The goal is to cover at least 75 percent of the country's baking quality wheat requirements. Production of coarse grains, notably barley and roughage, are to be increased enough to enable a decrease in grain imports one million tons below present levels by 1985. The area under oats will further decline and there will be no change for rye. The plan projected an average annual increase in the production of potatoes of 3.5 percent and sugar beets 3.0 percent based on 1976-1980 results. Potato production in 1984 was 11.9 million tons, 20 percent above the 1976-80 average. The GDR achieved a record yield of 24.3 tons per hectare, well above the 1981-83 average of 17.4. As a result, production went up 69 percent, the largest in the region. In 1985, the GDR planned to decrease area planted to potatoes by about 38,000 hectares or 8 percent. Production of sugarbeets came to 7.8 million tons in 1984, 11.5 percent more than the 1976-80 average. In 1985, the GDR planned to reduce planted area by some 4 percent.

There is no evidence, as of now, of intentions for making substantial changes in areas devoted to oilseeds. New varieties of rapeseed with lower euricic acid and higher per hectare yields are expected to increase output and allow a reduction in vegetable oil imports. Modest growth rates were set for all livestock inventories for the 1981-1985 plan period. Production growth is envisaged to come from better feeding efficiency, larger shares of roughage feeding and improved animal breeds.

Total meat production rose just 5.6 percent over the 1980-84 period to around 2 million tons. The largest gain in meat production occurred in 1984 in response to a significant rise in producer livestock prices. Increases were recorded in the output of each meat category allowing a 3.3 percent increase in per capita consumption. With an estimated 95 kilograms consumers in the GDR remained in first place for meat consumption in the COMECON bloc. Beef production is to be increased during the 1980s through beef type crosses. Simmenthaler and Charolais will be bred to the poorest milk cows.

For milk production the goal is a further increase in the average performance per cow, reaching 3,450-3,500 kilograms (at 4.0 percent butterfat) of milk a year by

1985. This is to be achieved by replacing low-performance with higher-performance animals through progress in breeding and feeding techniques.

The goals for pork production are to produce 1.6 to 1.69 million tons of pork per year (live-weight) and to raise 18.8 pigs per sow per year. Also, the average live weight should fall and lead to a higher slaughtering ratio as a result of more rational feeding.

Plans for sheep production were to increase the flock by 500,000 to 600,000 head to 2.6 million by 1985. The reasons for emphasis on sheep is to bring wool production to 7,000 tons annually, to use grazing areas more fully and to glean following harvest operations.

Broiler production is a growth sector in the GDR's livestock economy. The expansion of egg production has been moderated by the government which planned to bring output per hen to 210 by 1985.

Hungary

Agriculture is an important producing, supplying, and exporting sector of the Hungarian economy. Agricultural production contributed 14.7 percent to national income and employed 22 percent of the productive work force in 1983. Farm products provided 22.9 percent of total exports and represented only 9.3 percent of total imports in 1983.

Agricultural production was to grow 12 to 15 percent (2.3-2.8 percent a year) during 1981-1985 to insure some improvement in food supplies to the domestic market and to allow a boost in export (Table 7.1). This rate of growth is below the target (16 to 18 percent) set in the previous five-year period and may reflect the failure of agriculture to reach its past growth target. The target for 1981-1985 is, however, in line with the 15 percent (2.9 percent per year) growth rate achieved in the previous plan period. Agricultural investments in Hungary are characterized by strong fluctuations, accounting for 16.7 percent of the national total in 1981.

Crop production targets. Winter wheat and corn are the principal grains representing 90 percent of total output and 81 percent of the area under grain in 1984. These two grains together cover nearly 46 percent of the area under crops in Hungary. Barley is the third ranking grain with an area about one-fourth that of wheat. Both oats and rye have lost their significance. Corn is by far the principal feed grain, representing over 80 percent of feed grain output in 1984.

Targets for grain were set at 14.7 to 15.7 million tons by 1985 compared with the 1976-1980 average of 12.6 million tons (Appendix Table 7.1). The increase in production is to be achieved through higher yields. Wheat yields per hectare are to rise from the 1976-1980 average of 4.06 tons to 4.5-4.8 tons, and corn yields from 4.9 tons to 5.5-5.7 tons.

On the present indications, wheat and corn may be grown on a fifty-fifty basis on about 2.5 million hectares. There were, however, reports of plans for a drive to increase corn production gradually at the expense of wheat and barley.

Grain production in 1984 reached 15.7 million tons, a record. Sharply higher wheat production up to a record 7.3 million tons accounted for most of the gains in the 1984 outturn. Both wheat and corn yields at 5.41 tons and 6.09 tons per hectare in 1984 surpassed their target levels. Hungary aims to increase rice production in an effort to end expensive rice imports. The rice area was to be expanded to between 25,000 and 30,000 hectares by 1985 making use of low value land which can be utilized most effectively for rice production. This plan, however, did not materialize as the area planted to rice averaged only 13,000 hectares during 1981-84.

Oilseeds and animal meal production too are to be increased to slow the increase of imports of various protein feeds. Oilseeds occupy a relatively small position in Hungary's crop sector, accounting for only 6.6 percent of the country's arable land in recent years. Sunflowerseed is the major oilseed crop with a 1984 output of 640,000 tons obtained on 320,000 hectares. Rapeseed is the second most important oilseed crop with an 1984 output of 100,000 tons. Soybean production is small and has not come up to expectations of the planners.

Sugar beet acreage varied in the 1970s without a discernible trend. The growing area may decrease hand-in-hand with higher yields and development of varieties with higher sugar content. Production reached a high of 5.4 million tons in 1982. After a particularly poor harvest in 1983, production rose to 4.35 million tons in 1984. Sugarbeet production was planned to increase in 1985.

Livestock production targets. Livestock production grew faster than crop production between 1950 and 1979 and again in 1981 and 1983.

Meat production was projected to increase to 2.2 million tons by 1985, corresponding with a 42.8 percent growth over the 1980/81 level. This goal, in view of a 1.83 million ton output level in 1984, was not attained. In cattle production there will be a heavier concentration on specialized meat and milk production and a move away from dual-purpose animals. In the future more emphasis will be placed on specialized beef cattle production.

As regards milk production, the intent is to continue to raise the average yield per cow, now averaging 3,557 liters per year. Milk production reached 2.86

million tons in 1984, a 6 percent increase over 1981
which was 4 percent above 1980. Increased productivity
per cow, more than 20 percent in six years, caused most
of the change since milk cow numbers have trended downward since 1981.

Hog numbers increased sharply during 1983-84 but
declined 6 percent to 9.25 million head in 1985.
Unfavorable export markets accounted for this decline.
Pork production is to be expanded through improved feeding efficiency in large-scale industrial-type feeding
operations.

Poultry production for slaughter was to increase
from 367,000 tons in 1981 to over 500,000 tons by 1985,
making possible an increase of 16 to 17 percent of home
consumption and some 10 percent in exports. The trend
in poultry production is toward large-scale industrial
complexes. Poultry meat production stagnated at 415,000
tons in 1983-84, thus Hungary was not able to meet the
original target output. Production of eggs was to rise
from 4.4 billion eggs to 5 billion in 1985. This target
was apparently abandoned.

In sheep breeding, emphasis has shifted from wool to
meat to take advantage of renumerative foreign markets
for mutton, roasting lamb and sheep milk. Additionally,
there is a desire to increase self-sufficiency in wool.
The goal is to produce enough to cover half the needs of
the domestic clothing industry by the middle of the
decade. The plan envisaged a doubling in state purchases of sheep milk, bringing these purchases to five
million liters by 1985.

There is much room for improvement of feeding efficiency. Also there is a wide difference in feed conversion ratios between leading and average farms. Pig fattening, for example requires between 2.7 to 5.6 (an
average 4.2) grain units per unit of gain in live weight
which is about 20 percent more than in West Germany.
The production of one kilogram live weight gain for
slaughter cattle requires between 3.4 and 8.2 (an average of 5.0) grain units, and for poultry 2.6 kilograms
feed concentrate, about 25 percent more than in West
Germany. One of the reasons for these poor feed conversion ratios is the shortage of protein.

Poland

Agriculture remains an important sector of the
Polish economy contributing 18 percent to national
income and employing 28 percent of the labor force in
1982. Investments in agriculture amounted to nearly 20
percent of total investments in 1984, the second highest
ratio after the Soviet Union. Soil and climatic conditions only moderately favor agriculture.

According to the 1983-1985 plan, agricultural production was envisaged to be 9 percent higher in 1985
than it was in 1982. This means a 3 percent average
annual increase (Table 7.1). The principal objectives
of the next five-year plan (1986-90) are to balance
foreign trade in agricultural and food products, and to
increase supplies of food to the domestic market. In
both plan periods crop production is to receive greater
emphasis than livestock production to correct the imbalance between domestic food and feedgrain supplies and
consumption requirements.

Grain production was targeted to reach 24 million
tons in 1985 and 24.6-25.4 million tons in 1990 (Appendix Table 7.1). To attain this goal Poland must extend
her grain area from 8.1 million hectares to 8.6-8.7
million hectares and raise average yields from 2.6 tons
per hectare in 1982 to 2.9 tons in 1985. The 1984 grain
crop was estimated at a record 24.4 million tons, 24.6
percent higher than the average during 1976-1980.
Record wheat production and rye output accounted for
most of the record. Higher producer prices provided a
major impetus to larger grain output. The 1985 grain
crop is estimated at 23.9 million tons, 2 percent below
the previous year's bumper crop. Wheat production is
estimated to have reached 6.5 million tons, while coarse
grain production is estimated at 17.4 million, 6 percent
below the 1984 harvest. Potatoes and root crops are of
vital importance to the Polish livestock industries. In
recent years, better than half of the annual potato production has been used for feeding purposes. Lately
there has been a tendency to grow fewer potatoes. Production was targeted at 45.2 million tons for 1985 to be
scaled back to 44 million in 1990. Production totaled
37.4 million tons in 1984, thus the potato targets will
not be reached.

Production of sugar beets was to expand from the
1982 output of 15 million tons to 16.5 million tons in
1985 and to 17 million tons in 1990. Production reached
16.4 million tons in 1983, and 16 million tons in 1984
thus the 1985 target appears attainable.

Forage production has been the growth branch in the
crop sector. The areas under hay and feed roots in particular have shown a significant expansion. It remains
to be seen whether Poland is willing to cut fodder crop
production in order to obtain the added acreage needed
for increased grain production.

Rapeseed is Poland's major oilseed crop. Production
tends to fluctuate from year to year and has trended
downward since 1978. In an attempt to reduce dependence
on foreign meal, output goals for rape and other oilseeds were set at 910,000 tons in 1985 and at one
million tons in 1990. Poland's 1984 rapeseed output
reached 955,000 tons and plans call for further
increases in plantings.

Livestock and meat production goals. Livestock production including milk was sharply curtailed during 1981-83 because of reduced supplies of imported feeds. The January 1985 livestock census indicates no tapering off in the declining trend in Poland's cattle population. By contrast, recovery is indicated in Poland's hog and poultry numbers following three years of decline. The gains in hog numbers reflect improved feed supplies and more favorable procurement prices. The rise in poultry numbers is attributed to Government allocation of high-protein mixed feed to broiler enterprises. Still, hog and poultry numbers remained well below their peaks in the early 1980s. Plans called for cattle numbers to expand from 11 million head in 1983 to 12.5 million in 1985 and to 14.5 million in 1990. The cattle herd, however, numbered only 10.9 million head in 1985 and thus was below its target. Over the plan periods only a small increase in dairy cows was envisaged. In 1985 Poland's dairy herd should have reached 5.9 million head and remain at this size during the rest of the decade. This herd was expected to produce 17.5 million tons of milk by 1985 and 19.1 million tons by 1990. In 1984 a dairy herd of 5.7 million head produced 16.7 million tons of milk. Thus neither the targeted dairy herd size nor the milk output was realized in 1985.

The hog population was to reach 20.5 million in 1985 and 23 million in 1990. The 1985 target is only slightly above the 1976-1980 inventories. In 1985 Poland's hog population numbered 17.2 million head.

Large expansion in sheep breeding is anticipated with flocks expanding to 3.9 million head in 1985 and to 4.3 million in 1990. Both targets seem attainable.

Meat production was supposed to reach 3.4 million tons by 1985 and 3.9 million tons by 1990. In view of a 2.4 million ton output in 1984, Polish meat production fell significantly short of the 1985 target level. Pork is the leading meat category with output targeted at 2.2 million tons in 1985 and at 2.5 million tons in 1990. Yet, pork production in 1984 was down at 1.3 million tons reflecting the heard drawdowns of the previous two years. The goal for beef production was 1.25 million tons for 1985 and 1.55 million tons for 1990. In the light of a 770,000 ton output in 1985 even the 1990 target seems unrealistically high.

Poultry meat production is not expected to reach its 1981 output in the 1980s. Production for the rest of the decade is targeted at 300,000 tons, only about two-third the production recorded in 1981.

The plan calls for more inputs in agriculture and also asks the population to use efficiently all resources needed to overcome the present crisis. Polish agriculture is, at the present, seriously handicapped by shortages of fertilizers, plant protection chemicals and lack of machinery.[2] Fertilizer supplies to farms in 1984 were still 10 percent below the record 1975-76 use and supplies of plant protection chemicals were almost 50 percent below 1982 levels. While overall machinery supplies increased in 1984, shortages of batteries, tires, and spare parts for farm equipment remained a significant problem. Unless input shortages are overcome Polish agriculture will not have the potential to fulfill the planned targets. On present indications it is highly unlikely that the important underlying input supplies for success will be obtained. Optimistically, the plan assumes that economic reform will gradually increase production.

Romania

Agriculture is the most backward sector of Romania's economy and compares unfavorably with other countries in Eastern Europe. In 1983, agriculture contributed 19.7 percent to national income but employed 28.9 percent of total labor force. Agricultural output actually increased by 27 percent (4.9 per annum) during the past quinquennium instead of the planned 28-44 percent (Table 7.1). Much of the blame for the unsatisfactory performance is put on a combination of inadequate standards of farm management; excessive wastage during harvesting, transport and storage of produce; input deficiencies; technology lag; insufficient supplies of feedstuffs to the livestock sector and falling labor productivity.

For the 1981-1985 period, plans were to speed modernization in agriculture and to stimulate output so as to permit an increase in food consumption and to provide a surplus for export. Overall the plan called for an increase in agricultural output of up to 25-28 percent (4.5-5.0 percent per annum) over the 1980 level. The draft directives for the period 1986-90 are that agricultural production should be 29 percent higher than in 1985. A good deal of this growth is envisaged to come from improved labor productivity which was targeted to increase by 38.5 percent in the same period. Acceptance of the lower growth rate was not so much a case of lowered sights but the setting of more realistic goals for a change. Nonetheless, the goals remain very optimistic. Agriculture's 1981-1984 performance record would have to improve significantly even to attain the current scaled down objective. Planned targets for 1985 (6 to 6.8 percent) are hopelessly optimistic. Production shortfalls combined with stepped up exports to help pay the country's debt have created unprecedented food shortages which Romania has been suffering of late.

Balanced growth between the crop and livestock sectors is one of the major long-term goals for Romania's

agriculture. The plan called for the livestock branch
to hold a 45-46 percent share of total 1985 farm production, to be followed by a 50 percent contribution in
1990. The livestock share now runs 43-44 percent and
has altered very little in recent years. The crops'
contribution fluctuated with the fate of the harvest
season.

    <u>Crop production plans</u>. Grain production was tar-
geted to rise to 28.5 million tons per annum for 1981-
1985, a 47 percent increase over the output attained
during the previous five-year period (Appendix Table
7.1). In 1984 grain production reached a record 23.6
million tons. Mid-1980 targets for individual grains
are: corn, 16.5 million tons; wheat, 7.6 million tons
and barley, 4.1 million tons. The output of these
grains in 1984 were respectively, 12.6 million tons, 7.0
million tons and 3.5 million tons. In light of past
performance the production targets, particularly for
corn and barley are not only ambitious but also unreal-
istic. The main problems are low yields and the vulnerability of nearly half the country's grain growing area
(Danube basin) to floods and drought. Higher yields are
assigned a major role in the attainment of production
targets. Thus wheat and rye yields were expected to
increase from their 1982 level of 3.0 tons per hectare
to 3.5-3.6 tons per hectare and that of corn from 4.56
tons per hectare to 4.53-4.73 tons. Official statements
suggests that the grain acreage is to be expanded by 11
percent to around 7 million hectares. Grain crops are
afflicted with serious quality problems and subject to
harvesting and storage loses. Additionally, Romania
reports production on a bunker-weight basis without correction to standard moisture content. This reporting
practice tends to overstate availability of the crop by
about 15 to 25 percent.

    Besides grains, other key crops are sugar beets,
sunflowers, potatoes and vegetables. Sugarbeet output
was planned to reach 12.6 million tons compared to the
7.0 million tons output achieved in 1984. At the same
time, yields per hectare were forecast to increase from
25.6 tons to 38.5-40.0 tons. Production in 1983 fell to
4.8 million tons because of poor weather. Potato production, too, was to receive a powerful boost with output rising from the 1982 level of 5.1 million tons to
6.7 million tons. In 1984, Romania's potato output was
6.5 million tons. Romania would like to become more
self-sufficient in protein production to lessen its
dependency on imports to support a growing livestock
industry. Romania is Europe's largest producer of soybeans and the goal for 1981-1985 was an output of
835,000 tons, about two and half times the 1982 volume.
The largest soybean crop, 448,000 tons, was harvested in
1980. Sunflowerseed is Romania's primary oilseed crop.
The target was an output of 1.26 million tons against an

890,000 ton crop produced in 1984. Both targets appear more as a declaration of priorities than a production forecast.

Edible oil production was projected at 620,000-700,000 tons in 1985 compared with an output of 432,000 tons in 1982. Based on past production performance, this goal seems beyond Romania's reach. Output of vegetables was to climb by more than 20 percent.

Livestock production plans. In comparison with the targets set in the previous five-year plan period, the 1981-1985 plan targets show moderation, especially in the cattle sector. Cattle and pig numbers were planned to increase to 7.3 million head and 15 million head respectively. This compares with a January 1, 1985 cattle population of 7.0 million and a hog population of 14.8 million head respectively. According to U.S. Department of Agriculture analysts the reported cattle and hog numbers appear extremely optimistic in view of reports of severe health problems on livestock farms and ongoing meat shortages.[3] Since Romanian officials called for no increase in hog, beef and poultry numbers, both goals are too high to stand a chance of fulfillment.

The number of sheep and goats, which are important in Romania not only for meat but also for milk and wool, is expected to show a sharp increase to 19.0-19.5 million head. In 1985, Romania's sheep and goat population reached 18.7 million head but no change is expected for 1985. Achievement of plan target would have made Romania self-sufficient in wool.

Meat output was scheduled to reach 2.87-3 million tons by 1985, a 30 percent increase over 1976-1980 output. On a live weight basis that meat output target is 4 million tons. In 1984 meat production stood at around 1.82 million tons, and since the outlook is for no change, Romania will not achieve its planned target. Indeed, the 1985 goal seems on the optimistic side and divorced from reality.

Milk production was to come in for a larger share of attention than in the past. The 1985 output target, 8.9 million tons, compares with the 5.4 million ton output realized during 1976-1980. Actually, milk production in 1984 was below the 1980 level. Thus, the planned milk production target is also outside the realm of possibility. The 8.6 billion egg target, in view of the 1984 7.5 billion production level, cannot be achieved. To promote achievement of plan targets, funds representing nearly 20 percent of total investments in the national economy were allocated to agriculture in recent years.

Soviet Union

Agriculture remains a high priority sector in Soviet development during 1981-1985 and 1986-1990 as it still accounts for about one-fifth of the national income. Gross agricultural production for 1981-1985 is expected to be 12 to 14 percent above the average annual increase achieved during 1976-1980 and productivity in the agricutural sector is to increase by 22-24 percent. The production goal compares with the 14 to 17 percent growth target set for the 1976-1980 plan period and the actual 9 percent increase obtained. Productivity rose by 15 percent in the previous plan period. The 1981-1985 rate of growth is more in keeping with that obtained in 1971-1975 (13.3 percent overall growth rate (Table 7.1)). Actually, the planned levels of agricultural production have not been achieved during the last fifteen years. This suggests that the growth targets set for Soviet five-year plans are for the most part optimal goals for best-case scenarios. Besides the establishment of optimistic targets their unfulfilledness reflects infrastructural problems and technical weaknesses. Resources devoted to agriculture, especially production inputs have been insufficient, and incentives, managerial efficiency and farming practices have been inadequate. Inadequate supplies of the means of production such as fertilizer, plant protection products and farm machinery have been major contributors to output losses.[4] The attainment of 1981-1985 agricultural targets after the bad 1981-1982 and 1984 years are out of the question.

Crop production targets. In the grains area the main task is stated to be the laying of the foundation for self-sufficiency in grain. Average annual grain output was targeted at 238 to 243 million tons for 1981-1985, rising to 250-255 million tons in 1986-1990 (Appendix Table 7.1). The 1981-1985 target would be 33-38 million tons above the actual average harvest during the preceding quinquennium (205.0 million tons) and a considerably higher figure than in the two previous plans. The largest Soviet harvest record was 237.4 million tons in the 1978/79 season. As a result of the poor 1981-1985 crop outturns the grain production target was not attained. The target for individual grains together with their actual 1976- 1980 average production level are shown in Table 7.2.

By far the largest crop produced in the USSR is wheat. It accounts for roughly half of total grain output. As a result, the annual fluctuations in wheat output has a big effect on the totals for grain. The wheat production goal for 1981-85 was 100 million tons, practically the same as the 1976-80 average. Output has never reached the target level during the plan period. Wheat production in 1985 was estimated at 86 million tons, representing a rise of around 11 million tons over the previous year. The largest ever wheat crop was harvested in 1978 (120 million tons).

The 1981-85 plan for grains placed emphasis on coarse grains, mainly on barley, corn and pulses. There is new emphasis on corn production, with widespread soil improvement programs under way. Presently corn is primarily a fodder crop, most being fed to livestock as silage. Production of corn for grain rarely exceeded 10 million tons. While the 17 million ton target set for 1985 was not achieved the 20 million target for 1986-1990 will be difficult to reach unless plantings of high yielding varieties are sharply expanded.

Barley ranks second after wheat in the country's grain economy. Barley is favored for its adaptability to a variety of climatic conditions and its high yields. The combined barley-oats average annual production was planned to reach 87 million tons which would be roughly 15 million tons above the annual average harvest during the 1976-1980 period.

The 1981-1985 plan calls for average sugarbeet production of 100-103 million tons, an increase of 13 to 16 percent over the 1976-1980 average. Based on the 1981-1984 output performance of 74.8 million tons, this target has no chance of being fulfilled. Potato production was targeted at 89.1 million tons. Since output during 1981-84 fell considerably short of this target the 1981-85 output goal was not attained. Vegetable production is to increase to 29.4 million tons per year over the 1981-1985 period, and to 37 million tons during the 1986-1990 period. Output in 1982-1983 has already reached the 1981-1985 target.

Despite continuing government efforts, oilseed output has stagnated since 1971-1975 and there is relatively little prospect for any improvement in the near future. Cottonseed and sunflowerseed are the major oilseed crops in the Soviet Union. Cotton production has trended upward in the 1970s but further expansion of production in the 1980s can be expected to be limited by the practice of growing it in 100 percent irrigated areas. Cotton production declined between 1981 and 1984 to 8.6 million tons (seed cotton basis) and was about 14 percent below 1980's record harvest of 10 million tons. Cotton production fell short of its 1985 target of 9.2 - 9.3 million tons. Cottonseed production also declined to about 4.7 million tons by 1984. Cottonseed output is expected to approach 5.5 million tons by 1990. Sunflowerseed production was to reach 6.8 million tons a year during 1981-1985, and then rise to 7.2-7.5 million tons during 1986-1990. To attain the 1981-1985 goal would require an average yield of at least 1.28 tons per hectare and an area of 5.3 million hectares. In view of an 1981-84 output of 4.9 million tons, sunflowerseed production fell far short of the 1981-1985 target.

By 1985 vegetable oil production must be increased to 46 percent above the 1982 output of 2.6 million tons

Table 7.2 Grain Production Targets in the Soviet Union.

	1985 Plan	1976-1980 Annual Average	1971-1975 Annual Average
Wheat	100.0	99.7	88.9
Rye	15.7	10.9	11.5
Barley/oats	87.0	72.1	57.6
Corn	17.0	9.6	10.2
Millet	3.6	2.2	2.5
Buckwheat	1.6	0.9	0.9
Rice	3.1	2.3	1.8
Pulses	14.0	6.9	7.3
Other	1.0	0.9	0.9
Total Grain	243.0	205.5	181.6

to 3.8 million tons. Production, however, averaged only 2.6 million tons during 1981-84.

Livestock production targets. The most urgent task set for the 1981-1985 plan was further development of the livestock sector. Meat and livestock product output has not been able to meet the burgeoning demand forcing the Soviets to resort to import supplementation.

Average annual production of meat under the 1981-1985 plan was set at 17-17.5 million tons which is 2.1-2.6 million tons or 14 to 17 percent higher than the average output, 14.84 million tons, in 1976-1980. Targets for the 1986-1990 period call for average annual production of 20-20.5 million tons of meat. By 1990 beef and veal output should be as high as 9.5 million tons, pork 7-7.3 million, lamb and mutton 1.2-1.3 million and poultry meat 3.5-3.6 million. Total meat production in 1984 was about 17.0 million tons, of which beef and veal contributed 7.2 million, pork 5.9 million, poultry 2.8 million, and lamb and mutton 900,000. With larger livestock numbers on hand and Moscow's readiness to buy foreign grain and feeds to make up for shortfalls in domestic harvests, the lower limit of the 1985 meat production target is expected to be surpassed..

Over the longer term, the highest growth in meat production is scheduled for cattle, considering that there are large areas of natural pasture.

Soviet dairy farming is not satisfactory and plans call for further expansion both in terms of fluid milk and processed products.

Milk production should reach 97-99 million tons over the 1981-1985 period, compared with actual production of 92.7 million tons in 1976-1980. The task for the future is to achieve milk yields of at least 3,000 kilograms per cow. Milk output was about 97.9 million tons in 1984, thus the lower boundary of the target range has already been exceeded. Average annual milk production is to go up to 104-106 million tons during the 1986-1990 period.

Over the 1981-1985 period butter production was targeted to rise by 16 percent, and cheese production by 31 percent compared with the previous five-year period. Thus, butter production should reach about 1.48 million tons by 1985, and cheese production 850,000 tons. The production targets for 1986-90 were set at 1.57 million tons in for butter, and 1.08 million tons for cheese. Butter production rose to 1.56 million tons in 1984, and cheese production to 780,000 tons. Hence, butter production has already surpassed the 1990 target while cheese production in 1984 was roughly 65,000 tons short of the 1985 level. Beef production in the 1981-1985 period was to move more and more toward large-scale specialized feed-lot operations relying on concentrates and optimal feeding regimes. The production of pork has been traditionally highly centralized in large integrated complexes. Problems have resulted in the acquisition of feed supplies as the hog feeding operations were often remote from the sources of feed.

During the 1970s the Soviet Union expanded the poultry sector faster than any other sector of livestock farming. It remained an area again targeted for rapid growth in the 1981-1985 period. Production of eggs was to go up to 72 billion units per annum, by 1985 and to 78-79 billion by 1990. With egg production totaling 76.5 billion units in 1984, the Soviet Union was rapidly approaching its 1990 target. Considerable stress was also being laid on the development of the poultry meat sector which has contributed spectacular growth to Soviet meat production over the past five years.

Poultry meat output was to reach no less than 2.6 million tons in 1985 and 3.4-3.6 million tons in 1990. Special emphasis is to be placed on broiler production which was planned to rise from 700,000 tons in 1982 to about one million tons in 1985. The 1985 poultry meat target has already been met in 1984.

Wool production during 1981-1985 was to average 470,000-480,000 tons, or 10,000-20,000 tons more than the 1976-1980 average. Production in 1984 totaled 465,000 tons, thus the planned target will likely be fulfilled.

To promote the achievement of meat and livestock production targets, the plan called for a number of measures including; 1) increasing fodder production, 2) improving feeding efficiencies, 3) raising the level of mechanization in livestock production, 4) improving the livestock breeding herd, and 5) encouragement of private livestock production.

The new program for agriculture announced on May 24, 1982 signals a shift in Soviet livestock policy toward improving yields as a means of raising output. This would be a reversal of the policy emphasizing herd growth as the way to expanding supplies of livestock products.

A serious weakness of the Soviet livestock economy are low feed conversion ratios. The major source of feeding inefficiencies are unbalanced feed rations due to shortage of protein ingredients. Supplies of some feed additives are inadequate. Protein deficiency in feed rations was estimated to be 25 to 30 percent causing considerable wastage of and misuse of grains and feedstuffs. Nutritionally balanced feed rations could result in shorter fattening periods and better feed conversion ratios, and save an estimated 20 million tons of grains.

Further industrialization of the livestock sector remains the goal, with emphasis on joint (inter-farm) large, specialized animal and poultry farms. Their purpose is to achieve through concentration and specialization a higher output with lower costs and to increase labor productivity. In many cases these goals have not been achieved. At present, such complexes account for a relatively small share of livestock production, for example, 19.1 percent of pork production and 4.6 percent of beef and milk production. In support of raising the level of mechanization, a large share of new investment is earmarked for the construction of pig and poultry complexes, and to promote the use of modern technology in pig and poultry rearing. The livestock complexes are large, standardized, highly automated facilities developed to concentrate the breeding, raising, and feeding of livestock, including poultry.

Operation of private subsidiary farming is being encouraged in order to meet overall meat production targets. A favorable response by the private sector could mean a substantial boost to meat and other livestock product output. In order to facilitate achievement of targets, capital investments equal to 27 percent of total investments in the national economy during 1981-1985 were allocated to agriculture.

Country import projections by several independent sources are surveyed within the background of the previous discussion of official intentions and expected actual outcomes. The independent projections are summarized in Tables 7.3, 7.4, and 7.5. Unfortunately space does not permit a detailed description of the methodologies and assumptions employed in each of these projections. However, it is interesting and hopefully constructive to compare the wide range of forecasts of imports for the COMECON nations. Specific judgemental forecasts are ventured by this author that incorporate other studies and the consideration discussed above.

TABLE 7.3
Eastern Europe, Total Grains: Production, Consumption and Trade Projections 1980/81 and 1985/86.

	Bulgaria	Czech.	GDR	GDR[c]	Hungary	Poland	Romania	Eastern Europe	Eastern Europe[d]
1980/81									
Production	7.40	10.70	9.60	--	13.40	18.20	20.10	79.40	--
Consumption	8.00	11.50	13.70	--	12.60	26.30	21.00	93.60	--
Net Imports	-0.60	-0.80	-4.20	--	+1.00	-8.20	-1.10	-14.20	--
1985/86									
Production[a]	8.70	11.40	10.25	10.72	14.00	23.40	22.55	90.30	92.40;87.30
Production[b]	8.20	11.30	10.10	--	15.00	20.00	22.80	87.40	--
Consumption[a]	8.80	12.40	13.45	14.40	13.07	29.33	23.08	100.13	94.10;96.00
Consumption[b]	8.20	13.30	13.80	--	14.30	25.00	23.60	98.20	--
Net Imports[a]	-0.10	-1.00	-3.20	-3.68	+0.93	-5.93	-0.53	-9.83	-1.30;-8.40
Net Imports[b]	-0.40	-2.00	-4.00	--	+0.20	-5.00	-1.30	-12.50	--

[a]Source: Edward Cook, Prospects for U.S. Agricultural Exports to Eastern Europe Through 1985. ESS Staff Report No. AGESS810529, Washington, D.C., May 1981, p. 50.
[b]Schnittker Associates, Agra Europe, June 25, 1982, p. N/2.
[c]OECD, Prospects for Agricultural Production and Trade in Eastern Europe. Vol. 1, Paris 1981, p. 167.
[d]Tibor Barna, Agriculture Towards the Year 2000: Production and Trade in High-Income Countries. Sussex European Research Center, University Sussex 1979, pp. 115-117.

TABLE 7.4
COMECON: Protein Meal, Production, Consumption and Trade and Projections.

	1979/80[a]			1985/86[b]		
	Production	Consumption	Net Imports	Production	Consumption	Net Imports
	-------thousand tons-------					
Bulgaria	325	560	234	294.5	666	371.5
Czechoslovakia	116	801	685	92.5	900	807.5
GDR	123	1,123	1,000	153.8	1,432	1,278.2
Hungary	153	831	657	163.2	1,066	902.8
Poland	403	1,647	1,244	409.5	1,857	1,447.5
Romania	946	1,324	378	746.1	1,381	634.9
Eastern Europe	2,067	6,286	4,198	1,859.6	7,302	5,442.4
Soviet Union	5,184	5,725	1,404[c]	n.a.	8,600[d]	3,800

a U.S. Department of Agriculture, Foreign Agriculture Circular, Oilseeds and Products.
FOP 3-82, February, 1982, pp. 28-33. Production and consumption figures include sunflowerseed, soybeans, rapeseed, minor oilseeds meals and fishmeal.
b U.S. Department of Agriculture, Foreign Agriculture Circular, Oilseeds and Products.
FOP 4-82, March 1982, pp. 24-25. Net imports include imported meal plus meal from
imported materials.
c Edward Cook, Prospects for U.S. Agricultural Exports to Eastern Europe Through 1985.
ESS Staff Paper No. AGESS 810529, May 1981, p. 63.
d Anton F. Malish, Internal Policy, Decision Making and Food Import Demand in the Soviet
Union." A paper for the third meeting of the Trade Research Consortium 6/81, Table 5&6.

TABLE 7.5
Soviet Union: Total Grains, Wheat and Coarse Grains: Production, Consumption, and Trade and Projections, 1980/81 and 1985/86.

	1980-81			1985/86		
	Production	Consumption	Net Imports	Production	Consumption	Net Imports
			----million tons----			
Total Grains						
Malish[a]	178.7	217.2	33.5	235.0	251.0	18.0
Schnittker Associates[b]	--	--	--	212.0	240.0	30.0
Desai[c]	--	--	--	223.4;243.7	239.1;252.2	8.5;15.7
Bond and Green[d]	--	--	--	224.5	250.7	26.2
Barna[e]	--	--	--	220.3;210.0	225.6;222.5	5.6;11.8
Wharton Econometrics[f]	--	--	--	209.0	233.0	30.0
NFAC[g]	--	--	--	212.9	228.0	15.0;25.0
Johnson[h]	--	--	--	---	---	30.0;35.0
Wheat						
Malish[a]	98.2	116.7	15.5	112.0	114.0	4.0
Barna[e]	--	--	--	98.0;95.0	99.9;99.3	2.0;4.0
Coarse Grains						
Malish[a]	80.5	100.5	18.0	108.0	112.0	14.0
Barna[e]	--	--	--	119.0;112.0	122.7;120.4	4.0;8.0

a Anton F. Malish, Jr. "Internal Policy, Decision Making and Food Import Demand in the Soviet Union." A paper for the third meeting of the Trade Research Consortium, June 1981, Table 3.
b Schnittker Associates. Agra Europe, June 25, 1982, p. N/2.
c Padma Desai, Estimates of Soviet Grain Imports in 1980-85: Alternative Approaches. IFPRI, Research Report 22, Washington, D.C., February 1981, pp. 15 and 20. Assumptions: Grain production below average (223.4 million tons) and average (243.7 million tons); feed and seed and human and industrial = 227.9 million tons; waste = 5% for below average production and 10% for average production.
d Daniel L. Bond and Donald W. Green. "Prospects for Soviet Agriculture in the Eleventh Five-Year Plan: Econometric Analysis and Projections." Revised version, June, 1981. Nonpublished manuscript Table 10. Assumptions: Baseline projections; production = 1985 level and imports = 1981-85 average level.
e Tibor Barna. Agriculture Towards the Year 2000: Production and Trade in High Income Countries. Sussex European Papers No. 6, University of Sussex, 1979, p. 116.
f Wharton Econometric Forecasting Associates, Centrally Planned Economies Outlook, September 1982, p. 33.
g National Foreign Assessment Center, USSR: Long-Term Outlook for Grain Imports. A Research Paper ER79-10057, January 1979. Assumption: Gross production 238 million tons, minus 26 million ton waste.
h D. Gale Johnson. "The U.S., the Soviet Union, and the World Grain Economy." Paper presented at the 1980 Conference on U.S. Soviet Agricultural Relations. Grinnell College, Iowa, September 26-29, 1982. Proceedings, p. 38.

TRADE PROSPECTS

Bulgaria

Future trade patterns will be determined by the country's ability to achieve its production targets and consumption trends. The latter is of particular importance in the case of meat and other livestock products as the growth in consumption will affect the volume of feed imports and meat exports.

Comparison of the 3.4 percent annual growth rate target set for agricultural production with the 2.7 percent average growth realized in the 1981-84 period casts doubt on the attainment of the plan target.

Demand for protein feeds and coarse grains will continue to increase commensurate with the expansion of livestock production.

Bulgaria is not likely to be able to meet its grain production targets, especially for coarse grains. To achieve a grain output of 10.5-11 million tons would require an increase in the average yields per hectare from an average of 3.46 tons (1976-1980) to about 4.9 to 5.1 tons. Corn production is expected to fall short of domestic consumption requirements and imports will continue to be needed. Corn imports are estimated to vary between 250,000 and 500,000 tons (Table 7.3). The country will be an irregular and small importer of barley and other grains.

As in previous years domestically produced soybeans and sunflowerseed meals will not be sufficient to cover future consumptive needs. Hence, Bulgaria will continue to be a regular importer of protein meals. Future imports probably will change from soybean meal to soybeans to ensure optimum use of the country's processing facilities. Net protein meal imports may range between 250,000 tons and 400,000 tons (Table 7.4).

Additionally, Bulgaria needs to import cotton, hides and skins and tobacco. No significant change is expected in Bulgaria's cotton imports, most of which, as in the past, will come from the Soviet Union and to a lesser extent from Egypt. Bulgaria will also continue to rely on imports of live cattle, swine and frozen semen to upgrade the quality of its herds. Bulgaria's need for agricultural chemicals will continue to increase and part of the increase will have to be met from imports.

Bulgaria is a regular exporter of wheat and its exportable surpluses may be in the range of 300,000 to one million tons in the 1986-1990 period. Bulgaria is an important exporter of agricultural commodities. Major items are oriental tobacco, meat and meat products, cheese, furskins, aromatic plants and natural vegetable substances including essential oils.

Czechoslovakia

Czecloslovakia strives to become more self-reliant and keep imports to a minimum. The country regularly imported grain in the range of 780,000 to 1.4 million tons during the early 1980s the bulk of which were coarse grains, principally corn. Corn imports in this period varied between 0.5 to 1.1 million tons. Most of the remaining coarse grain imports are in the form of feed barley. Wheat imports were in the range of 220,000-260,000 tons. The volume of oilseed meal imports have ranged between 600,000-740,000 tons, about two-thirds of which was soybean meal.

Plans are to minimize future grain imports levels to 0.4-0.5 million tons to be supplied by COMECON. This target is substantially below Western estimates of probable import requirements (Table 7.3). It is hoped that a saving of one million tons of feed grains yearly can be achieved through improved feeding methods, and the use of grain substitutes. Because of the consecutive above average wheat harvests Czechoslovakia was able to cover its consumption requirements entirely from domestic supplies in 1984 and 1985. Barring a major harvest failure the intent is the virtual elimination of grain imports from the West to stem the outflow of hard currencies which is no longer affordable. Czechoslovakia generally attempts to purchase its import needs from fellow COMECON countries, turning to Western suppliers only when necessary. Czechoslovakia has a trade agreement with Hungary providing for the shipment of 200,000 tons of Hungarian corn annually to be compensated for in part by Czech exports of brewing barley and brewing technology.

Although plans are for increasing domestic resources of high-protein feeds, this will fall far short of requirements. Protein meal and oilseed imports into Czechoslovakia are expected to continue to expand resulting from the upgrading of protein content of mixed feeds and to meet additional requirements of the livestock sector. Officials now hope to keep such imports at the 0.7-0.8 million ton level, in line with purchases realized over the past plan period (Table 7.4).

Plans point to a move toward specialized dairy and beef production and away from current dual purpose herds. This will mean imports of breeding stock and a potential new market for U.S. breeding cattle.

Some reductions in imports of various foods are also being contemplated. Imports of certain vegetables (carrots, peas, onion) will be greatly reduced. Purchases of canned fruit will be slashed 40 to 50 percent. Fresh orchard fruit imports will be restricted to apples and peaches. Import of tropical and sub-tropical fruit will be frozen at recent levels.

Advanced technology on the basis of licenses is expected to play a part in agriculture's progress. The emphasis will be on providing seed stock and seedlings that complement the domestic range. As a means of achieving the planned targets for sugar exports, the plan calls for imports of high-quality seeds and modern farm machinery. Besides sugar, Czechoslovakia will likely maintain its traditional exports of hops, malt, beer and some meat. The country ranks third in world output and exports of hops and two-thirds of shipments abroad go to Western markets.

German Democratic Republic

The GDR already has a high degree of food self-sufficiency (85 percent) and this trend is expected to be further strengthened over the decade of the 1980s. In the case of livestock products, home production fully meets requirements, while in the case of products of plant origin, shortfalls have to be met from imports. The future supply situation for individual products varies greatly.

The GDR annually consumes between 12 and 13 million tons of grain, roughly one-third supplied by imports, mainly coarse grains. The GDR succeeded in attaining its grain production goals and intends to hold steady, or reduce, poultry and hog numbers while increasing those of cattle in the near term. Thus, future grain and pro- tein meal requirements will increase only moderately over current levels. Imports of grain, as in the 1970s, may fluctuate considerably in line with changes in domestic production. The volume of grain imports may be in the range of the 1976-1980 annual average of 3.4 million tons and four million tons per year (Table 7.3). Of this, coarse grains may account for about 3 million tons and wheat for some 500,000-600,000 tons. Corn will remain the principal coarse grain import aver- aging about 2 million tons. Malting barley will make up the bulk of other coarse grain imports.

Future import demand for coarse grains will continue to be affected by the size of potato crops. This is because a large percentage of the potato crop is normally fed as silage to swine and as roughage in beef rations. The percentage used as feed is determined by harvest results. It is the residual after setting aside requirements for food and industrial usage and seed. Additionally the growth in coarse grain imports may be reduced by the extent to which nongrain feeds, including roughages, are substituted for concentrates. Production of pelleted straw and straw-based semi-processed feeds are also likely to reduce demand for coarse grains. The

former is a new feedstuff with the addition of 2-3 percent urea. At present around two million tons of straw
are processed annually. It is planned that in future
years straw should meet about 10 percent of ruminant
feed energy requirements. This means that about three
million tons of straw would be used for feed purposes.

Wheat imports depend on the size and quality of
domestic output. The GDR needs annually about 1.4
million tons of milling quality wheat and 700,000 tons
of milling rye to meet its food requirements. It is not
able to produce all its milling quality wheat needed
domestically and future import requirements may remain
at the 1978/79-1980/81 range of 500,000-800,000 tons.
This may also include some feed quality wheat imports.
The GDR is attempting to replace part of the need for
hard wheats through the steaming of soft wheats. This
steaming process of soft wheats is to reach 600,000 tons
or nearly one-third of the bread wheats. Of total grain
import requirements COMECON suppliers, notably Hungary
and the Soviet Union, will be called upon to provide
about 1.0 million tons leaving the remainder to come
from the West.

The GDR has been importing large and growing amounts
of protein meals and cakes and a comparatively small
tonnage of oilseeds. Oilseed imports have varied widely
in recent years and generally trended downward. In 1983
the GDR's oilseed imports totaled 63,000 tons. During
1981-84 protein meal imports averaged 1.2 million tons.
Soybean meals have grown steadily in importance and have
accounted for about 90 percent of total meal imports in
recent years. The remaining oilseed meal is composed of
groundnut meal, cottonseed meal and sunflower meal. In
addition the GDR has, traditionally, been a significant
user of fish meal in livestock feeding rations. Imports
of fish meal, however, were on the decline in recent
years because of its scarcity and high price.

The GDR is expected to remain a substantial importer
of protein meals, particularly soybean meals, through
the 1980s. Imports may range from the 1976-1980 average
of 949,000 tons to 1.4 million tons through the remainder of the 1980s (Table 7.4). Fishmeal reportedly is
used widely in livestock feeding rations. The volume of
soybean meal imports thus may also hinge on changes in
fishmeal usage. Overall, GDR soybean meal imports will
be affected by the size of the rapeseed crop and by hard
currency reserves. To conserve hard currency the GDR
seems to seek barter and clearing account arrangements
for imports. In 1984 the GDR was the largest user of
soybean meal in Eastern Europe. In 1982 the GDR
arranged to exchange port equipment for 300,000 tons of
Brazilian soybean meal. Currently the GDR is able to
cover around 45 percent of its vegetable oil requirements from domestic oilseeds. Plans are to reduce vege-

table oil imports over the longterm. However, the GDR
will continue to import about 110,000 tons of vegetable
oils in the coming years.

With regard to future developments in the GDR's
livestock sector the situation differs from product to
product. The GDR is a net exporter of meat dominated by
pork. Pork, beef and veal exports are likely to be
maintained at their 1982-84 average level of 210,000
tons and 30,000 tons, respectively. Additionally, it is
expected that the GDR will continue to produce exportable surpluses in eggs, cheese and butter. It is probably that efforts will be made to eliminate the current
7,000-8,000 ton deficit in poultry meat.

The GDR is also an importer of canned and fresh
fruit and vegetables, rice, early potatoes, and hides
and skins.

Hungary

Agricultural exports provide benefits to the economy
through improvement in the balance-of-payments. Hungary
has a surplus in agricutural trade. The target was to
increase agricultural exports by one-third in 1981-1985,
with larger increases in hard currency (43 percent) than
in rubles. With a planned 2-3 percent annual increase
in agricultural production and a near stationary population, Hungary will assuredly have exportable surpluses.

The plan implies continuous austerity, the moderation of internal consumption by way of higher retail
prices and limitation on the growth of personal dispos-
able income. Slow growth in domestic demand should
enable Hungary not only to hold down food imports but
also to obtain additional export supplies. Pursuant
this goal authorities envisaged only moderate growth in
per capita meat consumption over the 1981-85 plan
period. Consumption of milk and dairy products over the
1981-1985 period was estimated to increase from the 1980
level of 162 kilograms to 175 kilograms, a rather over-
ambitious target.

Major export items are wheat, meat and pork specialty products notably canned ham, bacon and sausages,
live animals, cheese, fruit and vegetables and wine.
Also there is an increasing foreign demand for small
game animals. For certain commodities exports represent
a considerable proportion of total production. For
poultry it is 40 percent, live cattle and beef 45 percent, sheep 67.2 percent and for wheat 23 percent.

Hungary exportable grain surpluses are tied in with
the realization of planned grain and livestock production targets. The 14.7 to 15.7 million tons grain production goal because of the poor 1981 and 1983 harvests,
has not been fully achieved. Preliminary reports put
the 1985 harvest at 15.8 million tons which gives a 14.6

million ton production average for the plan period.
With a projected consumption of around 13 million tons
Hungary's 1985 exportable grain surpluses would approxi-
mate over 2 million tons. This would be double the
projected volume (Table 7.3). Additional grain could be
released for export should livestock feeders utilize
grain more efficiently in their operations. Hungary is
likely to continue its exports of feed concentrates and
mixed feed at the recent volume of 270,000 tons.

Growth of meat product export will be infuenced by
Hungary's ability to open up new markets in Western
Europe, the Middle East and the COMECON area. In this
respect much will also depend on Hungary's ability to
conclude supply contracts with Russia. Market assur-
ances would encourage the Hungarians to make additional
large capital expenditures needed for the modernization
and expansion of livestock production and processing
facilities. Should present market opportunities continue meat and meat exports may increase from the
347,000 ton volume in 1980 to between 400,000-430,000
tons annually in the remainder of the 1980s. The expansion of poultry exports is expected to slow down somewhat in the 1980s. Poultry exports may average around
160,000 tons in 1985-1990 compared with 186,000 tons in
1983. In the future, geese, ducks and guinea fowl are
expected to assume a greater importance in poultry
exports. Egg exports were targeted at 220 million eggs,
a significant underestimation of market potentials.
Exports already reached 350 million pieces in 1985.

The level of beef and veal exports will be determined by Hungary's ability to improve the efficiency in
beef production, the prospective demand by the world
market and Soviet Union imports. Without a substantial
increase in Soviet purchases, Hungarian beef and veal
exports may remain at their current level. Under favorable world market conditions beef and veal exports could
rise from the 1984 level of 50,000 tons to 55,000 tons
in the second half of the 1980s.

The outlook for Hungarian pork exports, in preserved
and canned form, is also affected by prospective world
market conditions and their competitiveness with East
and West European products and import restrictions in
the European Community markets. There is potential for
exports to increase from the 95,000 tons volume in 1981
to 150,000-160,000 tons in the remainder of the decade.
Favorable market conditions exist for mutton, lamb and
goat meat, and plans are to increase exports substantially above the 1981 level of 4,000 tons.

Overall, the Hungarian export targets were based on
the fulfillment of production plans, some of which, like
grain and meat, fell short of projected volumes. Also,
attainment of farm export goals has been affected by the
competitiveness of Hungarian products and the availability of imported protein feeds. To counter foreign com-

petition, the Hungarians will offer a broader range of products and intend to assume greater flexibility in meeting the changing demand on the export market. In an effort to enhance export capabilities and competitiveness Hungary has expressed interest in the import of meat processing equipment, equipment for producing dairy products, complete canning lines, refrigerating and deep-freezing equipment, and viticultural plants. Also, Hungary intends to explore cooperation possibilities with Western partners in turnkey installations of poultry and cattle farms. Under such projects Hungary would provide most of the basic plant and equipment; whereas the foreign partner would supply the livestock and feeds.

Despite plans for increased production of sunflowerseed and animal meal, Hungary will not be able to noticeably reduce its protein deficiency, at least not in the foreseeable future. Hence, its import requirements are not expected to vary much from the level of the past few years of 600,000-700,000 tons (Table 7.4). This would be in accord with recent Hungarian policy statements expressing the intention of stabilizing annual oilseed meal imports at the 1981 level of 570,000 tons.

Improvement programs for the upgrading of the sheep and beef herds as well as poultry stock is expected to continue, requiring the importation of breeding stock. Importation of live breeding cattle is not expected on a scale comparable to that of the 1970s. However, imports of a few breeding bulls and a substantial amount of Holstein-Friesien semen will likely be continued. Hungary will continue to import small amounts of meat, fruit and vegetables to supplement domestic supplies and to provide greater varieties.

Poland

To improve external balance must be considered the cornerstone of Poland's foreign trade strategy. Poland's net hard currency debt to the West was estimated at $26.8 billion at the end of 1984 and, with access to new loans limited, generating more hard currency through exports ranks high among planned priorities. Since 1982 Poland's overall trade balance improved and it managed to achieve growing surpluses. Poland's hard currency surplus in 1984 was $1.5 billion but experienced a deficit of 666 million rubles ($493 million) in intra-COMECON trade. Poland was also making good progress in balancing its agricultural trade; it's agricultural trade deficit was reduced from $2.4 billion in 1981 to $533 million in 1983 and the balance has improved further through 1985. Basically Poland's trade performance will hinge on the extent to which livestock, grain and feed production goals set for the remainder of

the 1980s are attained. The current austerity program
calls for a balanced agricultural trade. This is to be
achieved by curtailing imports of grain and feeds to the
minimum indispensable to complement domestic production
and by a slight increase in exports of Poland's staple
items.

Reducing grain and feed imports to 2.5 million tons
in 1985 is regarded as fundamental to the balancing of
trade. This, however, in view of existing supply and
consumption balances appears to be an elusive goal. In
1983 Poland imported 3.4 million tons of grains consisting of 2.4 million tons of wheat and the remainder of
feed grains, principally corn and barley. Protein meal
imports were only 446,000 tons in 1983. There continues to be a need of 2.5 million tons for milling
quality wheat.

Depending upon domestic grain and feed supply and
livestock production levels grain imports in the second
half of the 1980s could be in the range of three to four
million tons (Table 7.3) and protein meal imports
approximating 1.2-1.4 million tons (Table 7.4). Continued slow growth of livestock production, insufficient
foreign exchange, and above-average domestic grain production were putting a damper on 1985-86 grain imports.
However, with the continuation in the recovery in the
livestock sector, Polish grain imports should rise to
the projected level. Wheat will account for the bulk of
grain imports. Besides domestic grain production
levels, the volume of grain imports will also be affected by the size of fodder and potato crops. Because of
the decline in Polish meat supplies, rationing continued
and per capita consumption remained at 58.3 kilograms,
21 percent below the 1980 peak. In addition, domestic
demand has been depressed by relatively high retail
prices. The implications are that the policies favor a
balanced agricultural trade at the expense of domestic
meat consumption. Poland will continue to need to
import vegetable oils in approximately the same volume
as in previous years (125,000 tons).

Poland's prospects are not bright for realizing substantial expansion of traditional food and agricultural
exports. Meat and livestock exports face an uncertain
future because of the severe liquidation of livestock
numbers. Poland will likely remain a small net exporter
of meat. Beef and live cattle are expected to be exported in exchange for pork which, due to Polish consumers'
preference for it, is in short supply.

Poland, formerly a heavy butter and cheese exporter
became a net butter importer by the end of the 1970s.
Given today's difficulties, Poland is not expected to
resume its net butter exporter position until the end
the 1980s.

Growing efforts are being made to pay for imports of
food with exports of Polish industrial goods in order to

save hard currency for which other imports have to compete. The Polish Foreign Trade Ministry is reportedly
emphasizing the importance of barter trade.[5]

Romania

International transactions are projected to expand
at a faster rate than the economy as a whole, thereby
increasing Romania's dependence on foreign trade.
Foreign trade was supposed to increase by 50-57 percent
over the level reached in the 1976-1980 period.

The foreign trade plan envisaged that exports will
outpace imports permitting both the correction of trade
imbalance and a reduction of foreign indebtedness.
Specifically, exports were targeted to increase by 60-71
percent and imports 40-43 percent respectively. Stress
was placed on seeing the country's trade with developed
countries in balance by 1985 when imports were expected
to be $3.8-$5.3 billion and exports $3.8-$5.4 billion.
The deficit on trade with developing countries, largely
due to oil purchases, was intended to be scaled down to
$200-$600 million by 1985.

In an effort to alleviate its hard currency debt
burden, all industrial sectors have been instructed to
cut imports to the bone and to maximize exports in
1982. Consequently, the government has put pressure on
the agricultural sector to do its share in earning
foreign exchange and limiting imports. As a result of
these efforts Romaina's overall trade balances showed
marked improvement during 1981-84. The improvement came
despite deficits incurred in its agricultural trade.
Trade surpluses allowed Romania to cut its net hard currency debt from $9.8 billion in 1981 to $6.8 billion in
1984.

Romania was, until 1978, broadly self-sufficient in
grain, but has alternated between a net exporting and
net importing position during the 1981-85 plan period.
The prospective increases in production of feedgrains,
fodder and oilseed will be insufficient to support the
anticipated growth of the livestock industry. This will
necessitate some significant levels of imported feedstuffs during the 1980s, although purchases may be limited by Romania's continuing debt repayment difficulties
and hard currency shortage.

Net imports of all grains could range between 0.5-1
million tons in the second half of the 1980s (Table
7.3). This may consist of 0.5 million tons of feedgrains, primarily corn, some barley and other feedgrains. Corn meal is still an important element in the
Romanian diet. The country may also need to purchase
net, between 400,000 and 500,000 tons of wheat.
Requirements of wheat for human consumption are estimated at 3.5 million tons, while feed use of wheat may

range between one million tons and 1.5 million tons. An alternative to grain imports would be a cutback in livestock inventories and livestock exports.

In 1984 Romania imported 155,000 tons of oilseed meals and 375,000 tons of oilseeds, mainly soybeans. Romania's import dependence on soybean meal and soybeans is expected to continue at prevailing levels (Table 7.4).

To support its ambitious livestock programs, Romania will have to increase import of breeder animals. Other major agricultural import items are cotton and cattle hides.

Romanian plans to expand agribusiness will require the import of a wide range of machinery, equipment and technology. These include modern slaughtering, processing and refrigeration equipment; fertilizer spreaders and pesticide-herbicide sprays; harvesting machines for fruit, vegetables and forage; and cereal combines and tractor equipment. Also there is interest in purchasing systems for recycling animal wastes into feed and biogas production.

The 1981-1985 plan called for a 53.9 percent (9 percent per year) increase in agricultural exports. Romania derives about 50 percent of total hard currency earnings from exports of agricultural goods. It has indicated intention for stepped-up export efforts in all livestock sectors. Romania is the second largest meat exporter and the biggest exporter of eggs in Eastern Europe. Meat exports reached record volumes at around 220,000 tons in 1983-84 and plans are to increase them further in years ahead. This means attempts at increasing canned pork exports (North America and Western Europe), dressed meats (North Africa and the Middle East), baby beef (Italy, Middle East) and broilers (Middle East and Western Europe). The export of live animals especially sheep for slaughter (Middle East) is likely to be stepped up. Egg exports too should become increasingly important in 1981-1985 and later. Plans are to expand cheese and butter exports. Markets for Romanian cheese are being sought in Western Europe and the United States whereas butter markets are expected to be found in neighboring East European countries. There is also a desire to raise exports of fruit and vegetables.

Soviet Union

Foreign trade is an important factor in the development of the Soviet economy. Its rate of growth is much higher than that of material production. In the period 1981-1985 the volume of the Soviet Union's foreign trade turnover was planned to rise by 22.5 percent.

Soviet agricultural trade will be closely linked to developments in the field of production and consumption, and to resolution of transport and storage problems. It can be assumed that Soviet agricultural output will continue to vary with weather conditions, creating a large degree of instability in the country's economic performance.

The production records indicate that the USSR failed to reach the grain and oilseed production targets set for the 1981-85 plan. The Soviets apparently realized that these targets are out of reach as early as 1982 with the launching of the Food Program which shifted emphasis from 1985 to targets for 1986-90. Prospects for achieving grain and oilseed targets for 1986-90 are also subject to great uncertainty. This raises the question of the size of import needs and the regime's willingness to meet these needs.

The level of grain and feed requirements will be influenced by several factors. The first concerns prospective livestock numbers and progress in economies in feed consumption that could reduce the quantity of grain required per unit of output. Plans are for a wider adoption of balanced feed rations with respect to proteins, vitamins and trace elements. Feed grain requirements will also be affected by the production and quality of nongrain feeds such as hay, haylage, silage and feed roots.

It is becoming increasingly evident that the Soviet Union is lagging in its ability to produce enough oilseeds to meet domestic meal and oil requirements. Indications are that a major shift in Soviet livestock feeding methods is under way which is expected to further widen the gap between domestic oilseeds availabilities and consumption requirements. The level of protein meals, especially soybean meal, in livestock rations is expected to increase at the expense of grains. Hence, in light of Soviet livestock expansion plans, imports will continue to increase and make up an important share of total oilseed and oilseed products supplies.

In 1984 protein meal use amounted to about 6.2 million tons of which imports accounted for one million tons. It is projected that by the end of the 1980s, the Soviets may be feeding as much as 9.5 million tons of protein meal, of which probably 4 million would be imported.

Under the optimistic assumption that Soviet livestock and dairy products production targets remain at planned levels, and an average grain production of 250 million tons achieved, net annual grain imports may average in the range of 20-30 million tons during 1986-1990. U.S. Department of Agriculture's EE-USSR Branch projects Soviet net grain imports by 1990/91 at 20-25 million tons.[6] The composition of imports is likely to be affected by the wheat-coarse grain price relationship and the quality of domestic hard and durum wheats.

Admittedly the Soviet Union has the option to reduce feed grain and protein meal import requirements by buying large amounts of meat and dairy products from several suppliers. Despite the expected growth in the production of meat and dairy products, the Soviet Union will continue to import considerable quantities of beef, poultry meat and eggs. In 1984 it imported 925,000 tons of meat, consisting of 450,000 tons of beef, 225,000 tons of poultry, 150,000 tons of lamb and mutton, and 100,000 tons of pork. Egg imports amounted to 535 million pieces in 1984 and are forecast to stay at the same level in 1985. The Soviets imported 250,000 tons of butter in 1984, and are forecast to purchase about the same quantity in 1985. Soviet meat, butter and egg import requirements in the second half of the 1980s are likely to decline from present levels partly because of the modest per capita consumption goals and partly because of increased domestic supplies. Vegetable oil imports may stay close to their 1984 level of 786,000 tons. These imports covered nearly 23 percent of total vegetable oil consumption.

To facilitate increased imports the Soviet government has taken measures to expand and upgrade the throughput at ports and also to improve the system of internal distribution in order to expedite the movement of goods away from port areas. The expansion of grain handling capacity has eliminated the logistical limitations on grain imports existing in the 1970s. Soviet import capacity is now estimated at between 50 and 60 million tons of grain a year allowing the Soviets to handle all the grains they need to import in a single year.

SUMMARY AND CONCLUSIONS

Adding the annual targets for 1985 to the results for 1981-84 suggests that Czechoslovakia, the GDR, and Hungary are going to achieve their 1981-85 agricultural output targets, while the other COMECON countries will probably fail to achieve them. Inhibiting production growth are technical problems, a declining agricultural labor force, restrictive organizational systems, inadequate agricultural investments and shortage of the skilled technicians needed for continued agricultural mechanization and "industrialization." While these problems are common throughout the region, there are important differences among the individual states.

Bulgaria, Hungary and Romania are concentrating efforts on expanding their export potential while Czechoslovakia, the GDR and Poland stress rises in agricultural self-sufficiency. The Soviet Union aims at greater stability in year-to-year production.

An interesting addition to the list of COMECON priority tasks is the joint agricultural development within the region including support of the food and agricultural investments of the major exporters - Bulgaria, Hungary and Romania. The goal is still for each country to achieve its own food self-sufficiency.

COMECON countries' long-term grain and oilseed production outlook is for some gain in total output, but not enough to keep up with total consumption. This means that the region's grain imports are expected to decline by 1990 below mid-1980 import volumes. The combined net imports of the EE country members of COMECON may range between 4-5 million tons by the end of the decade compared with 14 million tons at the beginning of the 1980s. About half of this deficit may consist of wheat and the other half of coarse grains.

The key reasons for the decline in East European grains imports are; 1) maintenance of austerity programs which they have already begun and attendant slower growth in demand, 2) increased domestic production, 3) slower economic growth, and 4) foreign exchange constraints.

The area will remain an important or even an expanding market for protein feeds, purebred livestock, equipment for automated livestock feeding operations, and for machinery and equipment used in various branches of the food processing industry.

Imports of soybeans and soybean meal should increase as a result of continuing protein deficiencies and efforts to improve feeding efficiency. Domestic production of rapeseed and sunflowerseed will not appreciably reduce the regions' demand for soybeans and soybean meal imports. Net protein meal import requirements of the East European COMECON member countries may vary between 4 million tons and 5 million tons by the end of the 1980s.

Growth in livestock production is likely to be held at the pace of the 1981-1985 period. Production expansion in Hungary, Bulgaria, Poland and Romania will be tied to prospective export opportunities. Hungary, and Romania will remain net exporters of beef and veal while the rest of the member countries with the exception of the Soviet Union will be almost self-sufficient.

Pork will continue to be the most important meat and the COMECON region is expected to continue to increase its level of self-sufficiency in the 1980s. In poultry the region too will maintain its export surplus position with Hungary the chief source of supplies. Mutton and lamb are the least important types of meat in the COMECON, Bulgaria and Romania being the larger exporters.

In milk and dairy products each of the countries of the region are seen to remain self-sufficient. The export effort of the dairy industry will likely be confined to outlets in the Soviet Union.

On the whole, trade expansion in the second half of the 1980s is likely to be modest because; 1) agricultural plans call for increased levels of self-sufficiencies in food production; 2) there will be a slowdown in growth in livestock production from that observed in the 1970s and geared to export opportunities; 3) past efforts to hold down food prices may be difficult to maintain in light of budgetary constraints and balance of payments problems. Thus as food prices rise, consumption increases may slow; 4) shortages of hard currency will be compounded by large hard currency indebtedness; 5) there will be slower economic growth in the 1980s both in the East and in the West.

An important issue in COMECON-Western trade relations is the large hard currency indebtedness of COMECON nations. Net hard currency indebtedness of the East European countries to the developed market economies was estimated to have reached $60.3 billion at the end of 1981 and that of the Soviet Union $12.5 billion for a combined debt of about $73 billion (Table 7.6). Poland is the most heavily indebted of all socialist countries, followed by the Soviet Union, GDR and Romania in that order. Measured in terms of debt service as a share of exports, Poland, Bulgaria and the GDR are in the most difficult position.

The external adjustment policies adopted by the East European nations in response to rising debt burdens involved greater efforts and moderation in import demand growths. The upshot of these efforts was a turnaround in trade balances in East European countries, moving from deficit to surplus with the West in 1982 and apparently reducing their aggregate foreign debt. Net hard currency indebtedness decreased for all East European countries except Poland from their 1981 peaks to $50.8 billion in 1984. A build-down of the external debt position continues to figure explicitly in the plan targets of all East European countries. This will necessitate the achieving of continued balance of trade surpluses. Part of their general strategy to increase exports and reduce hard currency outlays will be increased emphasis on countertrade deals. The demand for countertrade will vary by countries.

NOTES

1. The relatively large labor force reflects the inclusion of workers employed in the food and fertilizers industries, agricultural technology and forestry as well as the low degree of mechanization of cultivation of the labor-intensive crops such as fruit, vegetables, wine and tobacco.

TABLE 7.6
Net Hard Currency Indebtedness of COMECON Countries 1971-1981[a].

Year	Bulgaria	Czech.	GDR	Hungary	Poland	Romania	Eastern Europe	Soviet Union	CMEA Banks[b]	Total COMECON
					(million dollars)					
1971	732	160	1,205	848	764	1,227	4,927	582	478	5,987
1972	909	176	1,229	1,055	1,150	1,204	5,723	555	1,240	7,518
1973	997	273	1,876	1,096	2,213	-1,495	7,590	1,166	1,454	10,570
1974	1,360	640	2,592	1,537	4,120	2,483	12,732	1,654	1,789	16,175
1975	2,257	827	3,548	2,195	7,381	2,449	18,657	7,451	2,790	28,898
1976	2,756	1,434	5,047	2,852	10,680	2,528	25,297	10,115	3,457	38,869
1977	3,169	2,121	6,159	4,491	13,532	3,388	32,860	11,230	4,154	48,244
1978	3,710	2,513	7,548	6,532	16,972	4,992	42,267	11,217	4,819	58,303
1979	3,700	3,070	8,950	7,300	21,500	6,700	51,220	12,100	3,000	66,320
1980	2,730	3,640	11,750	7,510	24,500	9,180	59,310	13,600	3,000	75,910
1981	2,135	3,755	12,640	7,900	24,250	10,350	61,030	12,500	n.a.	73,530

a Net debt equals gross debts to Western governments, commercial banks, suppliers, and other lenders less financial assets which consist of deposit placed with Western banks.
b International Investment Bank (IIB) and International Bank for Economic Cooperation.
Source: CIA, National Foreign Assessment Center, Estimating Soviet and East European Hard Currency Debt, ER80-10327, June 1980, p. 7. CIA, NFAC, Handbook of Economic Statistics 1982 A Reference Aid, CPAS82-10006. September 1982, p. 54 and p. 77.

2. As a result of inefficient harvesting, improper drying and storage, and processing almost four million tons of grain are estimated to have been lost every year.

3. U.S. Department of Agriculture. ERS. Eastern Europe Outlook and Situation Report. RS-85-7, Washington, D.C., June 1985, p. 15.

4. An elaboration on the scope of these production constraining factors is provided in D.G. Johnson and Karen Brooks, Prospects for Soviet Agriculture in the 1980s. Bloomington, Indiana University Press, 1983.

5. Polish containers have recently been swapped for tea, while machinery has been exchanged for Greek lemons. Rice and tea, mainly imported from China, is being paid for with mining equipment, factory machinery and other engineering products. See Agra Europe, July 19, 1985, p. N/2.

6. A.F. Malish "The Prospects for Agro-Business Development in the USSR." Remarks made at a discussion on Soviet agriculture organized by the Russian Research Center, Harvard University, February 6, 1984, p. 3.

REFERENCES

Agra Europe (London) Ltd, Agra Europe. (July 19, 1985): N/2.
Barna, Tibor. Agriculture Towards the Year 2000: Production and Trade in High-Income Countries. Sussex European Research Center, Papers No. 6, University of Sussex (1979): 115-117.
Bond, Daniel L., and Donald W. Green. "Prospects for Soviet Agriculture in the Eleventh Five-Year Plan: Econometric Analysis and Projections." Revised version (June 1981). Nonpublished manuscript Table 10.
Central Intelligence Agency, National Foreign Assessment Center, Estimating Soviet and East European Hard Currency Debt, ER80-10327 (June 1980): 7.
_____. National Foreign Assessment Center, Handbook of Economic Statistics 1982 A Reference Aid, CPAS82-10006. (September 1982): 54 and 77.
Cook, Edward, Prospects for U.S. Agricultural Exports to Eastern Europe Through 1985. ESS Staff Report No. AGESS810529, Washington, D.C. (May 1981): 50 and 63.
Desai, Padma. Estimates of Soviet Grain Imports in 1980-85: Alternative Approaches. IFPRI, Research Report 22, Washington, D.C. (February 1981): 15 and 20.
Johnson, D. Gale. "The U.S., the Soviet Union, and the World Grain Economy." Paper presented at the 1980 Conference on U.S. Soviet Agricultural Relations. Grinnell College, Iowa, Proceedings (September 26-29, 1982): 38.

Johnson, D. Gale and Karen Brooks, <u>Prospects for Soviet Agriculture in the 1980s.</u> Indiana University Press, Bloomington, 1983.

Malish, Anton F. "Internal Policy, Decision Making and Food Import Demand in the Soviet Union." A paper for the third meeting of the Trade Research Consortium (June 1981): Tables 3, 5 & 6.

Malish, Anton F. "The Prospects for Agro-Business Development in the USSR." Remarks made at a discussion on Soviet Agriculture organized by the Russian Research Center, Harvard University, February 6, 1984, p. 3.

National Foreign Assessment Center. <u>USSR: Long-Term Outlook for Grain Imports.</u> A Research Paper ER79-10057 (January 1979).

OECD. <u>Prospects for Agricultural Production and Trade in Eastern Europe.</u> Vol. 1, Paris (1981): 167.

Schnittker Associates. <u>Agra Europe.</u> (June 25, 1982): N/2.

UN ECE, <u>Economic Survey of Europe in 1984-1985.</u> New York, 1985.

U. S. Department of Agriculture. Foreign Agriculture Circular. <u>Oilseeds and Products.</u> FOP 3-82 (February 1982): 28-33.

U. S. D. A. ERS. <u>Eastern Europe Outlook and Situation Report.</u> RS-85-7. (June 1985): 15.

_____. Foreign Agriculture Circular. <u>Oilseeds and Products.</u> FOP 4-82 (March 1982): 24-25.

Wharton Econometric Forecasting Associates, <u>Centrally Planned Economies Outlook</u>. (September 1982): 33.

8

Doing Business with CPEs and Marketing Strategies for Western Exporters

Arvin R. Bunker, James R. Jones, and
Dennis M. Conley

Centralized control over the economy and handling of exports and imports through state controlled monopolies, in centrally planned economies (CPEs) confronts Western exporters with a different environment than trading with most other developed and many less developed countries. Many trade enhancement or restrictive devices used in market economies, such as tariffs and duties; quotas and quantitative restraints; special fees; credits; market development grants; and domestic production or marketing taxes or subsidies are often of less concern or consequence when dealing with CPEs. In their place, are complex administrative and planning procedures involving trade. Also major trade restrictions occur because of the adversary military or diplomatic environment which exists in varying degrees between the Soviet bloc and the West. This prevents trade in some products, increases costs and slows decision making for other products, and often makes trade subject to sudden and severe adjustments. Agricultural trade has often been subject to temporary but severe disruption. The U.S. led embargo of grain shipments to the Soviet Union following that country's invasion of Afganistan was joined by several other nations and was intended primarily to express political displeasure over actions by the Soviet Union.

Some trade restrictions or peculiarities grow out of differing concepts of political and economic planning and control between free market and centrally planned countries. Subordination of the economic system to the political system in CPEs usually affects the marketing environment. Control of political leaders over the economy means foreign trade organizations (FTOs) may pursue political rather than economic objectives at times. Planner preferences are consumer preferences.

Other trade peculiarities arise from the state's monopoly control over imports, which bestows significant economic leverage to the FTO over potential trading firms. On the other hand, this control also imposes substantial responsibility because trading mistakes may affect the entire country. Finally, some trade peculiarities result because rapid price changes in world markets are only slowly reflected in import decisions of CPEs due to administrative inertia. When adjustments are made they may be sudden and severe, but this depends explicitly on the decision to allocate foreign exchange.

Trading with CPEs is not necessarily inherently more difficult than trading with companies in market oriented countries. It is essential, however, from the perspective of an exporting firm or organization, that it understands the marketing structure and the trading preferences of the import decision makers it deals with.

STRUCTURE OF TRADING SYSTEM

The foreign-trade and internal distribution structure of a centrally planned economy is considerably different from that of most market economies. It was noted in Chapter One that since the system was devised initally in the Soviet Union during the Stalin period and later adopted by other Communist bloc countries there are many similarities in the foreign trade systems of these countries. As already noted, traders dealing with these countries cannot ignore the fact that business decisions are a reflection of a political process. Authority for economic planning in CPEs rests with the Party Presidium. The Party Presidium is selected from the Central Committee, which itself is elected from the Party Congress (see Appendix A in Chapter 1). Political and economic power is usually concentrated in the Presidium. Responsibility for defining and implementing economic plans goes to the Council of Ministers and from there to the State Planning Commission and to various ministries and their agencies.

The general model of import planning has overall economic policy and trade information flowing from the Party Presidium down to the lowest-level production agency. With overall trade goals provided by the presidium, lower level agencies prepare specific import and export plans and send them up through the system for approval by the Presidium. Several transfers of information from top to bottom and vice versa are made in estimating domestic production capability and subsequent foreign-trade needs. Lower level state enterprises, including state farms, cooperatives, manufacturing and distribution units, and privately held units (in some countries), have input by registering their resource needs in the production process. These needs are accumulated by the relevant ministry and evaluated in terms of achieving goals set by the Party Presidium.

When the resource base is insufficient to produce the commodities or goods needed, the relevant production ministry requests importing goods to meet the deficit. Decisions involve not only the Ministry supervising production but also the Ministries of Trade and Finance. These ministries must evaluate if foreign exchange is available and if import needs requested by the supervisory production ministry can be accomodated.

Import needs finally decided on are sent to the Party Presidium for approval. They may be approved or revised, depending on national policy considerations. When import needs are approved, FTOs, which are state trading agencies are usually in charge of securing those imports. These are the operational entities that exporters deal with.

Although the general trading structure is similar for CPEs, each country has unique situations and trading practices. It is important that prospective traders be well versed in the trading system of the specific country (Schmidt et al., pp. 44-48). The general agricultural trading apparatus by country is presented in Figures 8.1-8.6.

MARKETING STRATEGIES

Centralized economic planning systems present some unique business situations for those attempting to trade with these countries. Some transactions can become complex and may require services of specialists in EastWest trade. Once the exporter becomes familiar with specialized conditions in these countries, and ignoring political complications, most sales are no more difficult than those to non-centrally planned countries.

The almost complete government control over the economy generates special considerations traders should be familiar with. In most cases, companies in these countries that import or export are granted a monopoly for the entire country. Being a state monopoly trader can vest substantial purchasing or selling authority in a few individuals. A Western seller negotiating a sale is confronted with the realization of large potential future sales hinging on the success of the current sale. The seller may feel pressure to cut prices or offer extra services on the present sale to enhance prospects for future sales. The future buyer will likely be the same person as the current buyer.

On the other hand by being the only importer, mistakes in sales or procurement can have enormous consequences for the entire economy, at times even leading to political reactions. Import agencies are accountable for their actions and simultaneously subject to certain constraints and regulations imposed by higher echelons

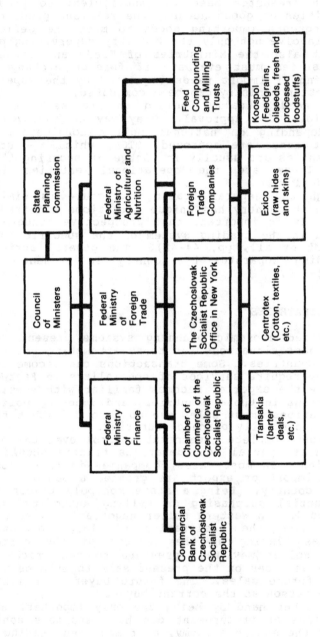

Figure 8.1 Agricultural Import Structure in the Czechoslovak Socialist Republic

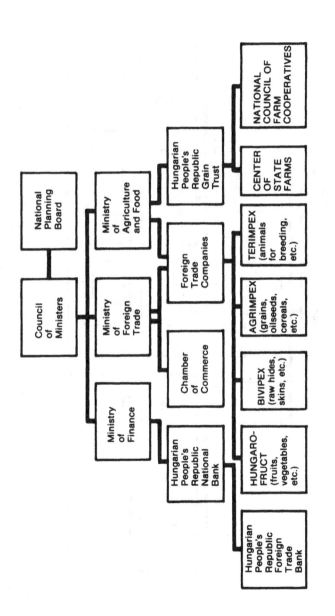

Figure 8.2   Agricultural Import Structure in the Hungarian People's Republic

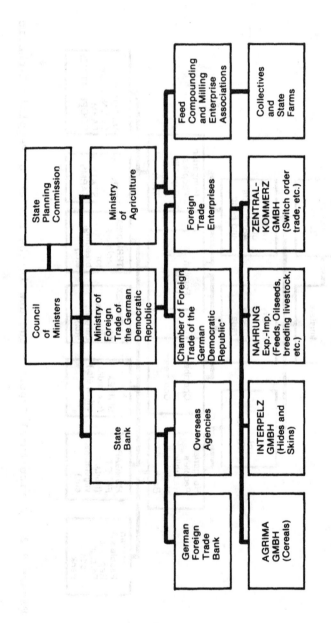

Figure 8.3   Agricultural Import Structure in the German Democratic Republic

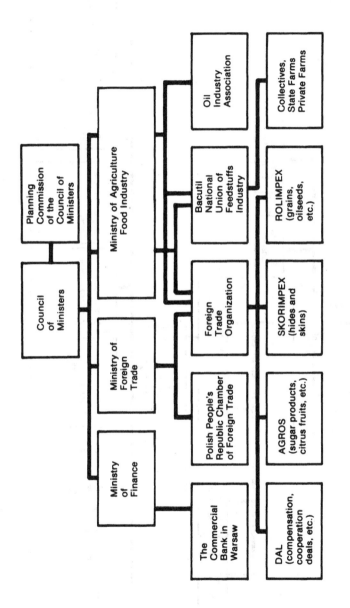

Figure 8.4   Agricultural Import Structure in the Polish People's Republic

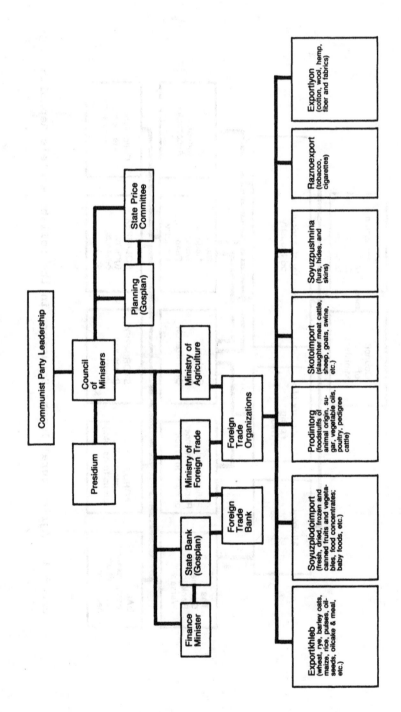

Figure 8.5   Agricultural Import Structure in the Union of Soviet Socialist Republics

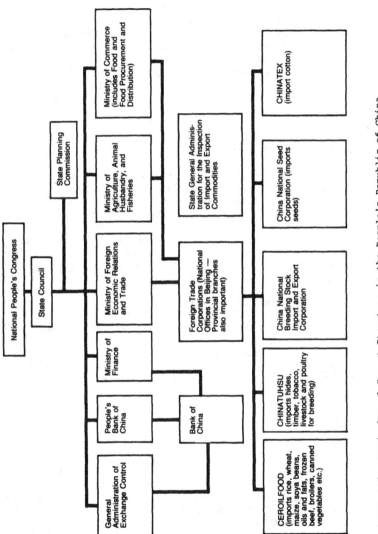

Figure 8.6 Agricultural Import Structure in the People's Republic of China

Sources: DePauw, 1981; National Council for U.S. China Trade, 1981, 1982; Department of Public Relations China Council for the Promotion of International Trade, 1983.

in the trading apparatus. Companies wishing to export to CPEs will need to select marketing techniques addressing traders' concerns and matching specific country requirements as perceived by the purchasing agency. As the contract which is representative of a business agreement between a grain exporter and the grain buying agency in Czechoslovakia in Appendix A illustrates, the transaction with an FTO entails usual commercial procedures that lay out certain agreed upon terms of sale. Commodity description and characteristics, quantity, price, delivery terms, method of payment, and supporting documents have to be worked out between the seller and the buyer just as they do in any other normal business transaction.

Presence in the Market

Because of the importance of agricultural imports in most countries and rather large annual purchases, import FTOs prefer to deal with suppliers who have an established presence in the market. Such a presence means an exporter can provide desired commodities in the proper quantity and of the preferred quality, along with related services, and can assure performance according to agreed upon terms. Several factors work together to establish a presence in the market in the eyes of FTO buyers. A high ranking in all factors is not necessarily essential to negotiate and successfully complete a particular sale. Overall success of a marketing program, however, will depend on the seller etablishing a presence and demonstrating its capability to meet the needs and concerns of buyers.

Direct source. A common attitude among Communist economic planners is that market intermediaries add more cost than value to a product and should be avoided in trading relationships. Economic decision makers in these countries frequently mention their desire for direct sources of agricultural commodities. Two major concerns seem to be behind this desire.

First is an assurance of supply. The need for assurance of a food supply is felt deeply in food-deficit countries. Concern over supply is especially great for principal food commodities such as grain and meat. In Poland, for example, meat shortages and abruptly increased food prices have contributed to political instability on several occasions. Grain is viewed as a principal food commodity and, as such, concern over supply reaches the highest levels of government. Inadequate or excessive grain imports by the FTO is likely to reach the attention of high state officials, who, in turn, react to possible demands by the people for adequate food supplies.

Aggravating the concern over inadequate food supply is the inflexibility of centrally planned economies to deal with the changing world food situation. Increases in grain imports, for example, to cover an unplanned domestic shortfall in production require not only grain purchases but also allocation of scarce foreign exchange, usually in hard currencies. Additional foreign exchange for grain imports must be withdrawn from previously planned use in other sectors of the economy, a difficult task for any government.

For these reasons, the prospect of stable food supplies for import, and relatively stable prices is a desirable factor for CPE food buyers. Trading directly with the source of these supplies rather than through intermediary firms might appeal to CPE importers. On the other hand, an organization directly representing a specific group of agricultural producers can be at a disadvantage relative to an intermediary multinational trading house. A company's willingness to trade products grown in countries around the world enhances the perception of CPE officials that the company can be a reliable supplier. Work stoppages by longshoremen or embargoes by one country can frequently be overcome by imports from another country.

A second reason behind a desire for direct sources is a perceived opportunity for reduced prices by eliminating intermediaries' profits. In conversations with CPE officials, especially those at higher levels, the idea of dealing directly is mentioned frequently. Traders in FTOs, however, mention this idea most often in the context of encouraging other companies to enter international food trade to provide additional competition and a check on multinational trading companies. The CPE traders deal with intermediaries every day and usually find services they provide justify the additional costs. Direct suppliers such as farmer cooperative export organizations are often encouraged by high level officials but will find that, in the end, they will have to be able to offer the same quality of marketing services as other trading firms.

Adequate Volume. As noted earlier, CPEs import substantial quantities of grains and oilseeds. Exclusive import rights means the FTO often may be in the market for relatively large purchases. Most sales contracts are for shipload quantities or more. The FTOs expect selling organizations with which they deal to be large enough to handle their requests for business. Also, importers are reluctant to deal with sellers that can supply only one-time or seasonal grain sales.

Assurance of Performance. Assurance that an exporting company can and will deliver according to agreed terms is a major consideration in selecting the supplying company. The most important evidence of performance is, of course, past activities in international trade. Eastern European trading officials stress the importance of a good performance record. Because selling to centrally planned countries requires some specialized knowledge and experience, one evidence of performance is to know a prospective export firm has traded success-fully with other centrally planned countries. Also as already noted, an important factor in evaluating poten-tial for performance is ability to originate grains from multiple origins, reducing the importing country's dependence on a single country or a few ports.

Convenient Access to Exporters. One principal handicap of U.S.-based grain exporters is the distance over which one must communicate to sell in Eastern Europe and Asia. The distance increases the cost of communications and makes it more difficult to get timely information due to differences in working hours. From New York there is a six hour time difference to Berlin, Prague, and Bucharest; a seven hour difference to Warsaw, Belgrade, and Budapest; an eight hour difference to Moscow; and a thirteen hour difference to Beijing. At best, normal business hours have only a two or three hour overlap, reducing opportunities for daily communications between buyers and sellers and hampering sales negotiations. Eastern European buyers strongly prefer to deal with sellers based in Europe. Poland, which uses a diplomatic mission in New York to gather information and negotiate with U.S. sellers is a notable exception. Traders in other countries generally prefer dealing with traders based near or in their country with sufficient authority to make sales decisions. Overseas offices in or near the importing country also increases opportunities for personal contact between buyers and sellers.

Market Information and Analysis. Rapidly changing world markets require traders continually to update their market information. The CPE grain traders prefer daily or more frequent contact with suppliers. Each importing FTO conducts, or has access to, economic analyses of the world grain situation. This is supple-mented by traders' frequent contacts with potential sellers. Multinational trading companies usually are viewed as excellent sources of market information and analysis. Their worldwide orientation closely matches preferences of CPE buyers, who usually abide by the principle that the FTO should buy from the lowest cost supplier, regardless of country of origin unless an explicit bilateral trade agreement requires trade with that country.

For Western traders, daily contact with CPE buyers is also useful. The CPEs are often large buyers of grain, and knowing, or guessing correctly, their purchase plans can create favorable trading positions for Western exporters in other countries.

Personal Acquaintance, Trust, and Responsibility.
Buyers for CPE countries emphasize a preference for
personal acquaintance with sellers. This emphasis is
related to assurance of performance stressed by FTO
buyers. Problems that may arise with negotiation and
delivery of a contract can be avoided or more easily
resolved between personal acquaintances. As with grain
traders in Western countries, those in CPEs have a
desire for honesty and trust in trade dealings. This is
especially critical in grain and oilseed transactions
where verbal commitments may trigger large market transactions long before written agreements are signed.

Pricing Considerations

It is commonly presumed that price plays a less
significant role in affecting decisions in centrally
planned than in market economies. Movements in world
price levels do not necessarily affect immediately the
quantities of goods produced and consumed within
centrally planned economies. However, firms selling to
centrally planned economies would be misled if they
assumed that the like- lihood is that total imports show
little response to world price movements and construe
this to mean it is not important to offer competitive
prices to FTO buyers. Foreign trade monopolies are
instructed to buy from the lowest bidder, other things
being equal, so an individual firm must be competitive.
Purchases from Western sellers cost CPEs hard currency,
and they are determined to conserve on outlays by buying
as cheaply as possible.

Occasionally, the lowest bidder will not get the
sale because some other consideration is involved, such
as possible willingness of another seller to offer
better nonprice terms. For example, willingness to
accept the CPE goods in a countertrade arrangement, a
reputation for superior quality and reliability, or provision of credit on desirable terms could result in a
purchase from other than the lowest bidder. These
exceptions are infrequent. Also instances occur where
politics or other noncommercial issues override price in
selecting import sources, especially where bilateral
agreements are involved with other governments or particular ideological matters make one seller more amenable
than another.

The mechanics of pricing when selling to FTO buyers
usually follow well-established procedures. Importers
in these countries use both formal public tenders and
private tenders to solicit bids from sellers. Pricing
terms of a sale are specified in a legal contract along
with other sale terms.[1] Flat price contracts, where

price is fixed at time of sale, is probably the most
common pricing arrangement for grain sales to CPEs.
However, some countries, Poland and the Soviet Union in
particular, sometimes will request basis or unpriced
contracts. In basis-price contracts, importers retain
the flexibility to fix the price any time before taking
delivery. The sales contract fixes the basis between
final price and a designated futures price. The pur-
chaser can then lock in a flat price at any time the
futures price is favorable by purchasing a corresponding
futures contract and turning its long futures position
over to the exporter. The flat price of the sale is
settled by adding the previously agreed-on basis price
to the price at which the futures are exchanged.

Financing and Credit

In an export-sales transaction, the seller prefers
payment as soon as possible, while the buyer prefers
payment after delivery or after resale. These conflic-
ting desires often are solved by extending credit.
Firms able to offer or arrange for credit increase
potential customers and sales. As noted in earlier
chapters foreign exchange shortages may mean trading
with CPEs requires financing, or otherwise trade is constrained. Credit may be extended for the short term,
usually defined as up to 180 days. Short-term credit
essentially covers working capital needs of buyer or
seller. Credit extended for more than 180 days usually
is considered long-term.

Credit may be extended to either buyer or seller.
Buyer credit usually is extended directly to the buyer,
usually the FTO, by the lending institution. The lending institution may be an exporting country, a Western
bank, an export-credit lending institution in the
exporting country, or the central bank of the importing
country. Buyer credit is available for short or long
terms and often for large amounts. Exporters extend
supplier credit to foreign importers. Supplier credit
includes letters of credit, cash against documents,
sales on open account, bills of exchange, and short or
medium-term credit from the supplier's own resources.

A major form of buyer credit to Eastern European
CPEs is available through the U.S. government. The
United States Department of Agriculture through the
Commodity Credit Corporation (CCC) offers export-credit
guarantees to commercial institutions for loans to
foreign purchasers of selected U.S. agricultural
products. The program is designed to expand sales of
agricultural commodities abroad by making financing
available to countries not otherwise able to afford purchases. The Export-Import Bank of the United States,

known as Eximbank, is an independent corporate agency of
the U.S. government that assists in financing U.S.
export trade and guaranteeing credits to overseas buyers
of U.S. goods and services. It guarantees and insures
shortand medium-term export transactions and export-debt
obligations held by commerical banks. In recent years,
the Eximbanks' role has been minor in financing
agricultural exports. Other U.S. government agencies
provide information and services to facilitate export
sales and underwrite risks for U.S. exports, including
Foreign Credit Insurance Association (FCIA), Private
Export Funding Corporation (PEFCO), Overseas Private
Investment Corporation (OPIC), as well as various
agencies of the Department of Commerce.

While governments or their financial institutions
may provide credit when the commercial banking system is
unable to, it is essential for exporters to develop a
working relationship with commercial banks. Commercial
banks can provide many financial services beneficial to
exporters and help a firm seek out alternative financing
sources and prepare necessary proposals. Banks often
initiate credit and assist with financing capital goods,
exports, and Eximbank transactions. Large commercial
banks also offer exporters various ancillary services,
including: buying and selling foreign exchange; collect-
ing foreign receivables; providing credit information of
foreign buyers; arranging introductions to foreign
banks; and supplying information on overseas taxes,
licenses, and regulations affecting foreign trade. The
bulk of U.S. export sales are short-term transactions
for which exporters seek commercial bank assistance.
The majority of this assistance is in documentary
letters of credit, with the remainder involving
collection of accounts, that is, drafts drawn on foreign
customers.

Although the financial systems are similar, the
particular situation of each country generates different
requirements for financing imports. In recent years,
credit requirements have varied widely from being essential for sales to Poland to only a minor factor in
negotiating sales to Czechoslovakia or the People's
Republic of China. The importance of credit to make
sales varies depending on the immediate credit needs of
the importing country.

Use of Marketing Intermediaries

Using marketing intermediaries is a possible strategy for developing and enhancing sales to CPEs. Several
types of intermediaries are common in international
trade, including export-commission agents, export
managers, brokers, foreign distributorships, and over-

seas offices. Of these types of intermediaries, only export commission agents, overseas offices, and possibly third party trading houses in countertrade transactions, are of much importance to agricultural trading. Brokers have little to offer, because there are few import decision makers and they are easily identified. Also, many CPE buyers do not favor dealing with brokers, feeling they add unnecessary costs to the product. The same feeling of adding unnecessary costs applies to export managers and foreign distributorships. The functions these intermediaries typically perform usually are provided by exporters or purchasing FTOs.

Foreign Trade Corporation officials interviewed by the authors generally have indicated that they prefer local agents or foreign-offices in their countries. The perception seemed to be that a local representative gives the FTO easier access to information and provides a nearby contact to solve problems arising from a sale. The local representative also can develop a working relationship with the FTO that could lead to improved estimates of import needs and timing of purchases. FTOs prefer a representative with authority to make sales decisions on the spot rather than one needing to clear decisions with an overseas office.

Several CPE countries have government sponsored agencies that will assist new exporters in making contact and conducting negotiations with importing FTOs. For most agricultural commodities, however, use of an importing country representative is not necessary, even for new entrants. In grain sales, especially, lower prices are more effective than agency contacts in obtaining sales. Agents in Western Europe are usually able to properly handle transactions for Western firms to Eastern European and Soviet Union FTOs.

The role of third party trading houses in countertrade transactions will be discussed in a later section of this paper.

Joint Ventures and Technical Assistance

The CPEs are frequently interested in joint ventures or technical-assistance projects. They believe such activities enhance technology and production in their countries and encourage companies or governments to offer these services. The actual sale or increase in sales, however, is usually not assured and may depend on the specific projects carried out. In general, FTO buyers do not favor tie-in sales and may not encourage or support a particular company in making such arrangements. However, if particular tie-in arrangements appear beneficial to higher officials in importing countries, FTOs will comply.

Joint Ventures. Past joint ventures between Western companies and CPE organizations for promoting agricultural trade include the following areas: feed milling; dairy, poultry, and cattle-feedlot operations; processing farm products into food; flour milling; baking; and others. The most common joint venture involves international marketing of products made in a CPE through a jointly owned enterprise. Both partners contribute capital and share in management decisions as well as profits and losses.

Investments in joint ventures are permitted in Hungary under special legal provisions. In addition, Bulgaria, Poland (before imposition of martial law), and Romania allow joint equity investment and joint management in such ventures. Yugoslavia is liberalizing legislation pertaining to joint ventures. In Hungary, it is possible to establish joint ventures that have legal characteristics similar to those in market-economy countries. They operate under co-management, co-ownership of capital, and sharing of profits and risk. Hungary has recently established an agency to promote and to assist in establishing joint ventures with Western partners. Most joint ventures exist in the industrial area. Several joint ventures are also being explored by Western agricultural firms with the People's Republic of China, especially in the areas of dairy and food processing.

Technical Assistance. Technical assistance is working with and through host country people to accomplish development goals. Purposes may be both long and short term and may change over time, even within the same country. Technical assistance involves building individual and institutional capability to deal with production, processing, or marketing problems in the receiving country.

Often CPE officials are unenthusiastic about joint ventures, but look favorably on technical-assistance projects -- perhaps as a result of favorable impressions of projects developed through the United States Department of Agriculture's Foreign Agricultural Service cooperator programs. CPE officials frequently compliment these projects for service to the agricultural sector in their countries. Most CPE officials avoid tying technical assistance to sales guarantees for specific U.S. companies. Officials in the ministries usually state that a company's participation in a technical assistance project would "enhance" trade opportunities, if trading terms were equivalent to terms offered by competing companies.

Traders at the FTOs may be less enthusiastic than at the Ministry level since technical assistance ventures

tied to trade impose restrictions on their decision
making abilities in purchasing products. FTOs do not
view their role as coordinating or tying in imports with
technical assistance projects. Usually, no agency is
specifically charged with this coordinating responsibility. Such a task would likely fall to the seller.
There are cases where technical assistance programs can
lead to increased or new orders for agricultural
exporters. If technical assistance programs of cooperator groups working through the United States Department
of Agriculture are viewed as successful and useful to
the host country, exporters may be able to use this
reservoir of good performance to propose additional
projects.

In summary CPE officials are interested in technical
assistance programs. However, they also adhere to the
principle that any trade should be economically competitive. Technical assistance increases visibility of a
company in the market by showing its good intentions but
this would not necessarily lead to sales contracts
unless the company was competitive with the lowest
bidder.

The FTOs have not coordinated purchases of agricultural products with technical assistance in the past.
Given the import structure of CPEs, it is difficult to
coordinate these two sides of a business relationship.
Agricultural commodity sales are short term and competitive, compared with technical assistance and joint ventures, which are long term, and the results less tangi-
ble. Under current operation methods, FTOs are autonomous in their decision making on when, where, from whom,
and at what price to buy agricultural commodities.

Role of Countertrade

Western firms doing business with CPEs will find it
necessary to become familiar with a variety of business
measures known as countertrade arrangements. As noted
in Chapter 1, these involve contractual arrangements in
which Western sales of goods are tied to reciprocal purchases of goods from the CPE in question. Countertrade
can be an important element of East-West trade. At
times, Western firms have found willingness to accept
countertrade commitments to be a prerequisite to making
a sale to a CPE partner. Apparently, this has not been
the case with many agricultural products, especially
grains and oilseeds but, depending on higher level
policy prescriptions, it may be a possibility in future
trade. Moreover, ability and willingness to do business
on such terms may present a competitive edge to an
export organization over other potential sellers. However, any party contemplating such a strategy needs to

be aware serious drawbacks to these arrangements frequently are encountered.

The CPEs stress countertrade in their commercial dealings with Western firms for short-term balance of payments and long-term market penetration reasons (Matheson, et al., 1977). They view countertrade as a means of generating or conserving hard currency through offset delivery provisions of countertrade contracts. The CPEs also recognize their own lack of success in penetrating Western markets and see countertrade as a way of using access and marketing capabilities of Western export firms in their own home markets to introduce CPE products to these markets.

Several forms of countertrade transactions exist (Verzariu, 1980). Counterpurchase is the form most likely to be relevant to agricultural export organizations, but there also are compensation, barter, and switch versions of countertrade. Counterpurchase is a form of countertrade involving counter deliveries of goods that are nonresultant products. The value of these goods is generally less than that of those sold by the Western firm. As a practical matter, this would almost have to be the case to be acceptable to an exporting company.

Compensation arrangements, also referred to as buyback arrangements, involve resultant goods directly derived from goods or technology provided by the West. Compensation arrangements have proved useful in East-West trade transactions where the Eastern country has abundant, cheap labor or raw materials, such as energy and other mineral resources, needed in the West, and the Western firm can provide technological assistance and capital equipment used in manufacturing products based on those inputs. This arrangement is seldom suitable for U. S. export organizations interested in selling raw agricultural materials.

Barter involves direct exchange of goods with no money being involved. This type of transaction is relatively rare. Switch transactions are based on multilateral (involving more than two countries) use of bilateral clearing accounts. These transactions primarily involve trading countries, usually other COMECON countries or less developed nations whose currencies are nonconvertible. Barter and switch transactions would be involved only in exceptional cases associated with cooperative grain-export transactions with CPE buyers.

Countertrade Procedures. Export organizations considering doing business with CPE buyers through a countertrade arrangement should be familiar with certain basic procedures. The complexity of countertrade precludes adequately describing all the details and nuances involved in such arrangements, but a few major procedures can be summarized.

A common and advisable procedure in countertrade deals is to draw up separate contracts covering export and import commitments. Also, a delivery clause probably would be necessary to take the original export party off the hook in the event the designated CPE supplier fails to meet delivery and/or service conditions on goods accepted as countertrade items. If, as is frequently the case, a third party, such as a trading house, is involved in disposing of goods accepted in countertrade, its role should be spelled out explicitly in the contract.

Because agricultural commodities, especially grains, oilseeds, and meats, are typically high enough on the import priority of most Eastern European countries for the foreign trade bank to set aside hard currency to meet the importing FTO's needs, legally binding countertrade contracts usually are not prerequisite to doing business. One alternative to a legally binding contract is to sign a letter of intent, which amounts to a gentlemen's agreement between the FTO and the Western exporter that the latter will purchase products from the country within a certain time period, provided they are available on suitable terms.

While the agreement is a part of the original sales contract, it is only morally, not legally, binding. This shows the exporter's willingness to reciprocate on a sale and apparently appeals to FTO negotiators at times, because they can demonstrate to their superiors they have negotiated a purchase that also may enhance the export efforts of their country. However, if such an agreement later is construed to be a charade in which the Western exporter had little intention of reciprocal purchases, it can damage seriously the reputation of the exporter. For this reason, it is important that exporters take such a letter of intent seriously. Indeed, because it is even more likely it will prove impossible to locate suitable goods to be purchased than under a legally binding countertrade agreement, the purchase is less likely to be carried out. This raises the danger CPE foreign trade authorities will become disgruntled with the firm and possibly even blacklist it in future transactions.

Locating products acceptable to an exporting organization as partial or complete payment for a sale to a CPE buyer is difficult because the exporting organization is seldom in the business of selling or using products imported. Developing the staff and organization necessary to do this may not be feasible. Firms frequently encounter this problem, and a fairly common way of getting around it is to use third-party trading houses specializing in disposing of goods acquired in countertrade transactions.

Numerous firms are concerned entirely with this type of trade or have special departments assigned to this task. Vienna, Austria, a center of East-West trade, and Hong Kong in the case of trade with China, harbor numerous trading houses specializing in countertrade. Several other locations, including Chicago, New York, and San Francisco serve as headquarters for firms engaging in such transactions. These organizations, for the most part, have no history of involvement in the grain trade, but rather focus on manufactured goods and raw materials. However, some multinational firms that have organized separate departments or subsidiaries to deal with grain sales are equipped for countertrade. The Swiss grain firm, Andre, created a subsidiary called Finco after World War II that specialized in barter, switch, and other arrangements. A U.S. firm, Phillips Brothers, Inc., which has entered the grain trade relatively recently, has an autonomous department specializing in countertrade. Japanese international trading companies also are at times involved in grain and countertrade transactions. More recently there has been considerable talk about newly formed U.S. export trading companies playing a role in countertrade transactions.

If a trading company is included in negotiations, a common procedure is for it to issue a "serious indication" document. This document shows willingness to assume the countertrade commitment of a specified commodity at a designated cost if certain specifications, clauses, and other conditions are followed. This permits the seller to proceed with negotiations, knowing the terms under which it can dispose of countertrade items.

If a third-party trading house is to be involved in locating countertrade goods, the contract should include a "transfer clause," specifying the countertrade commitment be transferred to it. So once countertrade products are located by the trading house, the purchase contract can be linked explicitly to the original export sales contract and countertrade commitment.

Relying on third-party trading houses to dispose of countertrade goods has drawbacks. Expenses for the services performed by these organizations become a cost of the original sales transaction. Also, it has been reported that involvement by trading houses is opposed at times by FTO negotiators who want to deal directly with the principal, because they harbor the impression that involvement of a third party increases the cost of imports. Another drawback is the seller's own negotiators are unable to master the intricacies of countertrade and are thus perhaps less proficient in conducting negotiations.

Trading houses willing to assume obligations to dispose of countertrade goods are not necessarily easy to locate. Few Western trading houses engage in countertrade as their principal line of business. They often find it more profitable to engage in the sales side of their clients' transactions and view assuming countertrade obligations as a necessary step in obtaining such accounts. In such cases, using trading houses as third-party intermediaries sometimes is more appropriate when the Western firm undertakes longer term commitments. See Appendix A of Verzariu (1984) for names and addresses of a few companies engaging in countertrade.

Another possible arrangement that might facilitate Western companies doing business with CPEs involves Western interests forming a counterpart or FTO marketing board to deal with CPE FTOs. Among its functions the organization could handle direct sales; countertrade; barter; switch deals; and cooperative arrangements, including technical assistance, joint ventures, and other trade matters. This would allow for pooling resources and developing specialized trading skills to meet the unique needs of the region and allow member cooperatives to spread the high cost of acquiring the expertise and facilities needed to carry out countertrade and other trading activities. It would expedite an export-import combination that could be useful in light of the hard-currency problems and bilateral trading tendencies of the CPEs. Finally, such a centralized organization might be in a better position to countervail any bargaining advantage held by FTOs who act as centralized buyers and sellers. The merits and drawbacks of such a centralized selling agency must be considered in the overall argument of whether the United States would benefit to have a marketing board similar to Canada's or Australia's and is only mentioned here in the limited context of countertrade. The larger issue of the pros and cons of marketing boards is beyond the scope of this discussion.

Pitfalls. Countertrade is not a guarantee to business success in CPEs. It is extremely difficult to orchestrate due to problems of arranging reciprocal purchases and inflexibilities of the planning bureaucracy in CPEs. The reciprocal feature is especially cumbersome because it requires a "double coincidence of wants" between buying and selling parties. Indeed, CPE officials engaged in buying grain and oilseed imports recognize countertrade can be wasteful and time consuming. Barring external pressure from higher authority, they would prefer only to give lip service to such schemes.

One of the more serious drawbacks to a countertrade strategy is that unsalable, out-of-date items often tend to get shuffled into countertrade. The CPEs often will not allow goods in high demand in Western markets to be used as linkage items for countertrade. Some firms have had an item they previously purchased to fulfill countertrade obligations suddenly disappear from the available list when the time to renew the annual contract came, because that product was no longer considered difficult to sell in hard-currency markets. If forced to take unwanted, inferior-quality products, exporters would have to sell them at a discount and absorb the discount as a surcharge to their own sales price.

Given low margins associated with grain, sales could be particularly difficult, unless the buying FTO made exception to its general rule of buying from the customer offering the lowest price. The FTOs charged with buying grain usually are unwilling to pay a premium to accommodate such countertrade arrangements. Nevertheless, Western firms selling non-agricultural goods to other FTOs have found them willing to buy from a firm at higher prices for payment in kind. This should not be ruled out completely in agricultural trade, because policies are sometimes fluid over time.

Countertrade arrangements face the potential of encountering import restrictions in the United States, particularly when imports reach relatively large dimensions, posing an additional difficulty. A case in point is the Soviet/Occidental Petroleum Corporation buy/back arrangement, which resulted in U. S. chemical producers lodging an injury complaint with the United States International Trade Commission. Occidental had entered a twenty-year agreement with the Soviet Union to provide construction assistance and superphosphoric acid to a Soviet plant and buy back up to 2.1 million metric tons of anhydrous ammonia annually and a million tons each of urea and potash to be marketed primarily in the United States. The commission initially voted favorably on the U.S. chemical producers' complaint and recommended quotas be imposed on imports of Soviet ammonia, starting with a ceiling of a million short tons in 1980. The President at first declined but later ruled favorably on the recommendation, although possibly more in response to the Soviet invasion of Afghanistan than the commission ruling. A later ruling by the International Trade Commission failed to support the contention of domestic injury, so the temporary import quota was removed. This type of complication always is a possibility when a firm assumes a role involving imports as part of an export transaction.

Possible Countertrade Goods. As noted, the most difficult aspect of countertrade is identifying goods that could feasibly be acquired as a part of the reciprocal arrangement. One possible avenue of conducting countertrade would be to arrange to market commodities acquired in countertrade deals through U.S.

farm supply companies.

Because farm supply companies could provide the CPE coveted direct access to the U.S. market, U.S. export organizations might be able to exploit successfully a countertrade strategy by disposing of items received in countertrade through these organizations.

Success is of course not assured for such a venture. It remains to be determined if CPEs are willing or able to provide these goods as partial payment for agricultural imports. Also, the rigidity and complexity of dealing with state run FTOs could be particularly pronounced, because goods indicated as potentially desirable imports are handled by FTOs supervised by ministries other than the agricultural ministry that supervises import needs for grains, oilseeds, and other agricultural goods. Such linkage arrangements are not uncommon in countertrade, and in certain cases, specialized FTOs have been set up by the CPEs to facilitate these arrangements. Nevertheless, to coordinate separate import/export organizations, not just in the CPE but in the United States as well, would be a formidable task entailing considerable resources and risks.

It is just as difficult to identify goods CPE markets might be willing and able to offer in countertrade as it is to identify what goods U.S. exporters would be able to acquire. Current examples of countertrade transactions only give limited insight into this question. Table 8.1 includes a list of some countertrade arrangements successfully organized in Eastern Europe, involving types of goods that might be of interest to U.S. companies. These examples do not serve as ideal prototypes a U.S. exporter could emulate easily, because they usually do not directly involve agricultural commodities in the transaction.

In certain cases, CPEs have availed themselves of Western technology through licensing agreements. Thus, their technology may be comparable to Western standards, simply beause it is Western technology, though not necessarily of the same level and vintage. In one example of countertrade, the British firm, Massey-Ferguson-Perkins, licensed a tractor model to be manufactured at the Ursus tractor factory in Warsaw. Full capacity of 75,000 tractors per year is scheduled under the license. In another arrangement, Machinenfabrik Gebruder Claes, a West German firm, signed a cooperation agreement with the Hungarian FTO, Komplex, covering joint production amd marketing of eight-row machines for corn picking and husking. Also International Harvester (United States) has a cooperative agreement with the Stalowa Wola factory in Poland and has taken delivery of crawler tractors built for the U.S. market.

Some countries have been allowing transportation equipment and services to be counted as fulfilling

TABLE 8.1   Countertrade Arrangements in Eastern Europe

Importing Country	Western Supplier	Signed	Import	Export
Poland	Massey Ferguson	1974	Equipment for Ursus tractor plant	Diesel engine and tractors
Poland	Rhone-Poulenc Institut Francis du Petrole	1975	Chemical products and textile fibers	Sulfur
Poland	Creusot-Loire	1976	Equipment and technology for fertilizer plant	Fertilizer
Poland	Katy Industries	1976	Machinery and working programs for shoe production	Shoes
Poland	Krupp-led consortium	1976	Coal gasification plants	Ammonia, urea and methanol
Hungary	Stieger	1974	Licenses and equipment for manufacture of tractors	Tractor axles
Hungary	Steiger	1976	Technology and components for tractor manufacture	Tractor axles
Hungary	Semparit	1976	License for tire production	Tires
Hungary	Levi Strauss	1977	Material (under negotiation)	Levis
Hungary	Bekoto Pertersime Pvba	1977	Technology for production of egg-collecting and incubator vehicles	Egg-collecting and incubator vehicles

TABLE 8.1 (continued)

Importing Country	Western Supplier	Signed	Import	Export
Hungary	Hesston	1977	Harvestors (80) and hay-handling systems (12)	Heads and gearboxes for Hesston's harvestors
Hungary	Machinenfabrik Gebruder Claes GmbH	1978	Agricultural equipment	Agricultural equipment
Hungary	Holsten Brauerie	1984	Licensing beer production	Malt
Hungary	Philip Morris	1984	Licensing cigarette production	Tobacco, printed material
Hungary	Benetton	1985	Fashion and sports clothing	Sawn wood
Hungary	APV International	1985	Milk and beverage processing equipment	Consumer ready beverages
G.D.R.	Berlin Consult	1975	Construction of meat-processing plant	Meat
G.D.R.	Chemie Linz	1976	Pesticides, herbicidal agents, and fertilizers	Potassium salts and special chemicals
G.D.R.	Dow Chemical	1976	Chemicals	Metalworking products, plastics, and chemicals
G.D.R.	Vereinigte Edelstahlwerke	1977	Fine steel products	Potash fertilizers
G.D.R.	Kloeckner Industrie	1978	Potash granulation plant	Potash granulates and unspecified products
Yugo-slavia	DeCloet	1985	Tobacco seed and technical assistance	Raw tobacco

U.S.S.R.	Cooperative Whole-sale Society	1984	Wool, mohair clothing, footwear	Raw materials for pharmaceuticals
U.S.S.R.	Coop Suisse	1984	Clothing, shoes	Tea, sauerkraut, honey, marmalade, vegetables
U.S.S.R.	Cooperative Forbun-det Skogsallians Nordik Agro	1984	Clothing, shoes	Fruits, vegetables, fish
U.S.S.R.	Scancoop Coopera-tive Society, Huhtamaki, Kauko-makkinat, Suneva	1984	Textiles, clothing, shoes, perfumes, paints	Greases, jams, fruits, vegetables
U.S.S.R.	Konsum, Kraus & Co. Dauber, Ledex Handelsges	1984	Clothing, shoes, leather goods	Rawhide, vegetables, honey, jams
U.S.S.R.	Marathon	1984	Consumer goods	Canned jellies, jams
U.S.S.R.	Peru	1985	Various manufactured goods	Fishmeal, minerals

Source: Pompiliu Verzariu, Countertrade Practices in East Europe, the Soviet Union and China: An Introductory Guide to Business; op. cit. pp. 79-86; Jenelle Matheson, Paul McCarthy, and Steven Flanders, "Countertrade Practices in Eastern Europe," East European Economics Post-Helsinki, pp. 1305-1311.

countertrade commitments. Poland, in particular, has
achieved considerable success in expanding its shipyards
and fleet of ocean vessels. So this area may offer a
possibility for negotiating long-term arrangements
involving chartering vessels or purchasing vessels for
use by cooperatives for shipping exports to third markets. Hungary also has allowed freight services to
third countries to be counted as countertrade services.
A great deal of study would be required before feasibil-
ity of such an arrangement could be assessed.

SUMMARY

Many trade regulations market economies use to
control imports and exports are not used in centrally
planned economies. In their place, however, is a complex set of administrative procedures and regulations
built into the state trading aparatus. While significantly different than regulations in market economies,
these trade regulations do allow Western traders access
to markets and the potential for profitable trading
relationships.

Most trade regulations by centrally planned economies grow out of subordination of the economic to the
political system and the resulting need for centralized
control over the economy. Other trade restrictions are
imposed by Western governments because of the military
and diplomatic adversary relations between centrally
planned economies and Western countries. These trade
restrictions prevent trade in some products and make
trade in others subject to severe disruption.

Centralized planning for domestic production, consumption, and trade is conducted annually and involves
all levels of the economy. Ultimate authority, however,
is vested in the Communist Party Presidium. Once import
needs are identified, foreign trade organizations are
directed to secure the needed goods. Agricultural commodities are procured from the least costly source,
other things equal.

The most important unequalizing factor to price is
availability of credit. Hard currency shortages means
centrally planned economies must often secure credit for
purchases. Government to government credit is often
preferred. Such government credit may eliminate price
competition from other countries, but not from other
sellers in the United States.

Despite complex administrative restrictions to
trade, once Western exporters become familiar with these
restrictions, trade is no more difficult operationally
than trade with market economy countries. Traders
wishing to trade agricultural products with these

countries should be aware of several principles that guide FTO buyers.

Because of the importance of agricultural imports and rather large purchases, foreign trade organization buyers prefer to deal with suppliers that have an established market presence. Factors contributing to an established market presence include: direct source of product, preferably from more than one production area; ability to supply large quantities; past experience in trading with centrally planned economies; convenient access to exporters; availability of market information and analysis; and personal acquaintance, trust and responsibility.

Within a given planning period, changes in world prices for agricultural commodities are seldom reflected in prices to end users. Consequently, quantities imported may be unresponsive to price changes depending on foreign exchange allocation decisions. Companies selling to foreign trade organizations, however, quickly discover that for any given sale price this is usually the determining factor in vendor selection.

Marketing intermediaries located in centrally planned countries are seldom necessary when selling agricultural commodities. If adding their services increases the price of the commodity, this will usually result in the company not being the low bidder. Most transactions can be adequately handled from offices in Western Europe.

Few joint ventures between Western companies and centrally planned economies have been established. Some centrally planned economies encourage joint ventures that use Western technologies to produce goods to be exported to less developed countries. To date there are few of these in food industries.

Technical assistance projects sponsored by the United States Department of Agriculture have been favorably received by most CPEs. However, it is difficult to persuade foreign trade organizations to commit to purchases of agricultural commodities from a particular company because of support of a technical assistance project.

Although countertrade is becoming much more common throughout the world, it has been seldom used in trade of bulk agricultural commodities. High-level ministers and state planning officials enthusiastically support counter trade principles, but foreign trade companies and traders in centrally planned countries producing or using goods to be traded are less enthusiastic. In fact, many avoid the complications and restrictions countertrade imposes on their operations. Nevertheless, Western firms willing to engage in countertrade may find a significant trading advantage when dealing with cen-

trally planned countries. There are a number of pitfalls to avoid, however, and careful study of the pros
and cons should be undertaken before embarking on this
course of action.

In conclusion doing business with centrally planned
economies requires the seller to become familiar with
procedures and regulations unique to the specific market
as in any other situation. The vagaries of international diplomacy can shift the trade winds in unpredict-
able fashion and there are certain idiosyncracies
associated with state trading, but otherwise many, if
not most, commercial principles apply when consumating
transactions with CPE buyers that apply to private
parties.

NOTES

1. For more details on procedures involved in
transacting and pricing grain sales, see Nielson
Conklin, et. al. (1979).

REFERENCES

Clark, Christopher M., Kathryn Dewenter, China Business
    Manual 1981. Washington D.C.: The National Council
    for U.S.-China Trade (1981).
Conklin, Nielson, Gerhard Wilbert, Reynold Dahl "Pricing
    of Grain Exports and the Role of Futures Markets."
    Minnesota Agricultural Economics. No. 614 (December
    1979).
Department of Public Relations, China Council for the
    Promotion of International Trade. China's Foreign
    Trade Corporations and Organizations: A Directory.
    Hong Kong: Evergreen Pub. Co. (1983).
DePauw, John W. U.S.- Chinese Trade Negotiations. New
    York: Praeger Pub. (1981).
Gardner, H. Stephen. Soviet Foreign Trade: The Decision
    Process. Boston: Kluwer-Nijhoff Pub. Co. (1983).
Hillman, Jimmye S. "Nontariff Barriers: Major Problem in
    Agricultural Trade." American Journal of Agricul-
    tural Economics, 60:3 (August 1978): 491-501.
Holzman, Franklyn D. International Trade Under Communism.
    New York: Basic Books Inc. (1976).
Matheson, Janelle, Paul McCarthy and Steven Flanders.
    "Countertrade Practices in Eastern Europe." East
    European Economies Post Helsinki. Congress of the
    United States, Joint Economic Committee (August 25,
    1977).
Posthumus, Dale and William Huth. "Selling the USSR: A
    Primer for Exporters." Foreign Agriculture. Vol
    XXIII No. 2 (February 1985): 4-8.

Schmidt, S. C., J. R. Jones, D. M. Conley, A. R. Bunker. Cooperative Grain Trade Opportunities in Eastern Europe. U.S. Department of Agriculture, Agricultural Cooperative Service, Research Report 21 (May 1984).

Verzariu, Pompiliu. Countertrade Practices in East Europe, the Soviet Union, and China: An Introductory Guide to Business. U.S. Department of Commerce, International Trade Administration (April 1980).

Verzariu, Pompiliu, International Countertrade: A Guide for Managers and Executives. U.S. Department of Commerce, International Trade Administration (November 1984).

Appendix A
Standard Contract Issued by Koospol
Stating Terms to Be Met by
Grain Exporter

RE: General Contractual Conditions on Yellow Corn/Hard
      Winter Wheat - Basis C and F

Seller	:	
Buyer	:	Koospol a.s., Leninova 178, Praha 6
Commodity	:	No. 3 US Yellow Corn - max. 15% moisture
Quality	:	Final at loading according to US Grain Inspection Certificate, US Phytosanitary Certificate and Veterinary Certificate.
Quantity	:	.....metric tons 5% more or less in sellers option at contract price, delivered weight without prorata.
Price	:	US $..../per metric ton.
Delivery Period	:	Delivery in Hamburg from .... to .... both dates included.
Parity	:	C and F Hamburg free out berth Rethe-Speicher. In case vessel will discharge at another berth than Rethe-Speicher seller will reimburse to the buyer US % 2, .../mt.
Packing	:	In bulk
Payment	:	At first presentation of documents with Ceskoslovenska obchodni banka a.s., Praha. The documents must be remarked: "Final destination Czechoslovakia."

```
Documents
Must         :  1/  Commercial invoice 5 copies
Contain         2/  2/3 Bill of Lading
                3/  Copy of registered airmail letter to
                    E. Clemens, Borgfelder Strasse 34,
                4/  Copy of Charter Party
                5/  Copy of Phytosanitary Certificate
                6/  Certificate of Origin
                7/  Weight Certificate
                8/  Inspection Certificate
                9/  Veterinary Certificate
Discharge    :  Demurrage/Despatch as per Charter Party.
                Daily Discharge rate 4.000 mt wwdshex

Insurance    :  To be covered by buyers.

Arbitration :  At Gafta in London under the arbitration
               rules No. 125 of which both parties admit
               to have knowledge.
```

All other terms and conditions as per Gafta contracts No. 27/30.

Contributors

Josef C. Brada, Professor
Department of Economics
Arizona State University
Tempe, Arizona 85281

Arvin R. Bunker, Agricultural Economist
Agricultural Cooperative Service
U.S. Department of Agriculture
Washington, D.C. 20005

Dennis M. Conley, Industry Analyst
Corporate Strategy and Research
Farmland Industries, Inc.
Kansas City, Missouri 64100

Gail L. Cramer, Professor
Department of Agricultural Economics
    and Economics
Montana State University
Bozeman, Montana 59715

Joel R. Hamilton, Professor
Department of Agricultural Economics
University of Idaho
Moscow, Idaho 83843

James R. Jones, Professor
Department of Agricultural Economics
University of Idaho
Moscow, Idaho 83843

256

Bob F. Jones, Professor
Department of Agricultural Economics
Purdue University
West Lafayette, Indiana 47906

C.S. Kim, Agricultural Economist
International Economics Division
Economic Research Service
U.S. Department of Agriculture
Washington, D.C. 20005

Hassan Mohammadi, Ph.D. Candidate
Department of Economics
Washington State University
Pullman, Washington 99164

Stephen C. Schmidt, Professor
Department of Agricultural Economics
University of Illinois
Urbana, Illinois 61801

C. Peter Timmer, Professor
Harvard University
Cambridge, Massachusetts 02138

Alan J. Webb, Agricultural Economist
International Economics Division
Economic Research Service
U.S. Department of Agriculture
Washington, D.C. 20005

Kenneth B. Young
Winrock International
   Livestock Research Center
Morrilton, Arkansas 72110